PRINCETON AERONAUTICAL
PAPERBACKS

1. LIQUID PROPELLANT ROCKETS
David Altman, James M. Carter, S. S. Penner, Martin Summerfield.
High Temperature Equilibrium, Expansion Processes, Combustion
of Liquid Propellants, The Liquid
Propellant Rocket Engine.
196 pages. $2.95

2. SOLID PROPELLANT ROCKETS
Clayton Huggett, C. E. Bartley and Mark M. Mills.
Combustion of Solid Propellants, Solid Propellant Rockets.
176 pages. $2.45

3. GASDYNAMIC DISCONTINUITIES
Wallace D. Hayes. 76 pages. $1.45

4. SMALL PERTURBATION THEORY
W. R. Sears. 72 pages. $1.45

5. HIGHER APPROXIMATIONS IN
AERODYNAMIC THEORY. M. J. Lighthill.
156 pages. $1.95

6. HIGH SPEED WING THEORY
Robert T. Jones and Doris Cohen.
248 pages. $2.95

PRINCETON UNIVERSITY PRESS · PRINCETON, N. J.

NUMBER 6

PRINCETON AERONAUTICAL
PAPERBACKS

COLEMAN duP. DONALDSON, GENERAL EDITOR

HIGH SPEED
WING THEORY

BY ROBERT T. JONES AND DORIS COHEN

PRINCETON, NEW JERSEY
PRINCETON UNIVERSITY PRESS
1960

HIGH SPEED AERODYNAMICS

AND JET PROPULSION

———————•◆•———————

BOARD OF EDITORS

PRINCETON, NEW JERSEY
PRINCETON UNIVERSITY PRESS

PREFACE

The favorable response of many engineers and scientists throughout the world to those volumes of the Princeton Series on High Speed Aerodynamics and Jet Propulsion that have already been published has been most gratifying to those of us who have labored to accomplish its completion. As must happen in gathering together a large number of separate contributions from many authors, the general editor's task is brightened occasionally by the receipt of a particularly outstanding manuscript. The receipt of such a manuscript for inclusion in the Princeton Series was always an event which, while extremely gratifying to the editors in one respect, was nevertheless, in certain particular cases, a cause of some concern. In the case of some outstanding manuscripts, namely those which seemed to form a complete and self-sufficient entity within themselves, it seemed a shame to restrict their distribution by their inclusion in one of the large and hence expensive volumes of the Princeton Series.

In the last year or so, both Princeton University Press, as publishers of the Princeton Series, and I, as General Editor, have received many enquiries from persons engaged in research and from professors at some of our leading universities concerning the possibility of making available at paperback prices certain portions of the original series. Among those who actively campaigned for a wider distribution of certain portions of the Princeton Series, special mention should be made of Professor Irving Glassman of Princeton University, who made a number of helpful suggestions concerning those portions of the Series which might be of use to students were the material available at a lower price.

In answer to this demand for a wider distribution of certain portions of the Princeton Series, and because it was felt desirable to introduce the Series to a wider audience, the present Princeton Aeronautical Paperbacks series has been launched. This series will make available in small paperbacked volumes those portions of the larger Princeton Series which it is felt will be most useful to both students and research engineers. It should be pointed out that these paperbacks constitute but a very small part of the original series, the first seven published volumes of which have averaged more than 750 pages per volume.

For the sake of economy, these small books have been prepared by direct reproduction of the text from the original Princeton Series, and no attempt has been made to provide introductory material or to eliminate cross references to other portions of the original volumes. It is hoped that these editorial omissions will be more than offset by the utility and quality of the individual contributions themselves.

Coleman duP. Donaldson, General Editor

PUBLISHER'S NOTE: Other articles from later volumes of the clothbound series, *High Speed Aerodynamics and Jet Propulsion*, may be issued in similar paperback form upon completion of the original series in 1961.

CONTENTS

SECTION A

AERODYNAMICS OF WINGS AT HIGH SPEEDS

ROBERT T. JONES

DORIS COHEN

CHAPTER 1. FUNDAMENTAL CONSIDERATIONS IN THE DEVELOPMENT OF WINGS FOR HIGH SPEEDS

A,1. Review of Wing Theory for Low Speeds.

FUNCTION OF THE WING. The chief advantage of air transportation over other forms of travel is the great speed that can be achieved with a relatively moderate cost in terms of energy, or fuel expended per mile of flight. An airplane of efficient aerodynamic form may have a drag less than one twentieth of its weight. The energy required in steady flight is therefore less than one twentieth of the weight times the distance flown. By proper adjustment of the wing loading and altitude this value may be maintained up to speeds approaching the speed of sound. Beyond the speed of sound the efficiency of flight tends to be reduced—but it seems probable that advances in aerodynamics and in jet propulsion can overcome the loss, at least for moderately supersonic speeds.

It is easily seen that the efficiency of the airplane is the result of the favorable aerodynamic properties of the wing. If one contemplates travel by a rocket, which overcomes gravity and achieves its distance solely by virtue of the kinetic energy imparted at the beginning of the motion, then it is found that the energy requirement is much greater—of the order of the whole weight times the distance. The energy expenditure of a wingless rocket or projectile thus corresponds to a lift-drag ratio of about one, a figure which can be surpassed easily by almost any form of winged body.

EARLY EXPLANATIONS OF WING FLOWS. The possibility of the favorable aerodynamic properties of airfoils was not apparent to early students of hydrodynamics and, in fact, prior to actual demonstrations of mechanical flight, solutions of the hydrodynamic equations were used as the basis of predictions that such flight would prove impossible. This situation is perhaps not surprising when it is recalled that the equations of hydro-dynamics or gas dynamics do not provide a definite solution to the problem of the airflow around a wing, but, in fact, furnish a multiplicity of possible solutions. The identification of the proper physical solution in every case requires additional information not contained in the equations of motion nor in readily determined boundary conditions.

Helmholtz' solution. An early attempt to explain the action of wings on the basis of hydrodynamic theory is that of Helmholtz [1]. Helmholtz adopted as a representative shape an inclined flat plate. The simplest

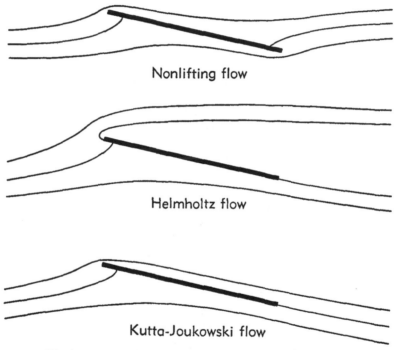

Nonlifting flow

Helmholtz flow

Kutta-Joukowski flow

Fig. A,1a. Hydrodynamic solutions for the flow over a flat plate.

mathematical solution in this case depicts a flow without lift (Fig. A,1a, top) which involves infinite velocities around the edges of the plate. To avoid this physically unrealistic situation, Helmholtz devised a solution in which the streamlines leave the leading and trailing edges tangentially, forming external surfaces of discontinuity which enclose a body of "dead" air above the wing (Fig. A,1a, middle). Such a configuration is certainly not contrary to experience. However, when the aerodynamic forces associated with the Helmholtz flow are compared with those that can be achieved experimentally with properly designed wing shapes, the Helmholtz flow is seen to give relatively small lift and an excessive amount of drag. The lift force developed is

$$L = \frac{\pi}{4} \alpha \rho_\infty U^2 c \tag{1-1}$$

or, in the usual coefficient form,

$$C_L = \frac{\pi}{2} \alpha \tag{1-2}$$

for small angles of attack α. Since the resultant force is at right angles to the plate, a drag equal to the product of the lift and the angle of attack is developed.

Kutta-Joukowski flow. A more suitable starting point for the description of wing flows was found later by Kutta and Joukowski. Kutta [2] prescribed Helmholtz' condition of tangential flow at the trailing edge, but not at the leading edge (Fig. A,1a, bottom). In Joukowski's solution [3], the airfoil was rounded to permit flow around the leading edge without separation, and was terminated in a cusp so that Kutta's condition could be applied to determine a unique value of the lift for each angle of attack. The Kutta-Joukowski flow consists of two independent solutions of the equations of motion, one determined directly by the normal components of velocity of the solid surface and the other—the circulatory flow—involving only a parallel sliding motion around the surface. The entire lift depends on the secondary flow, which cannot be determined directly from the normal boundary conditions, but only indirectly from the condition of tangential velocity at the trailing edge. For thin flat sections at small angles the lift developed is, in coefficient form,

$$C_L = 2\pi\alpha \tag{1-3}$$

and in two-dimensional frictionless motion the drag is zero.

PHYSICAL CHOICE OF BOUNDARY CONDITIONS.

Effect of the boundary layer. The occurrence of the Kutta-Joukowski type of flow depends physically on the action of the boundary layer, which causes the flow to leave the trailing edge tangentially. Separation of the flow around the leading edge can be avoided for a small range of angles of attack because of the thinness of the boundary layer in this region. The essential difference between a leading edge and a trailing edge is evident in the case of a flow through a pipe. Fig. A,1b shows schematically two examples of streamlines provided by the perfect-fluid theory for such a flow. However, the boundary layer formed inside the pipe will prevent the turning of the streamlines at the exit, so that the actual flow will resemble that depicted in the bottom part of the figure, rather than in the top part.

Leading and trailing edge conditions. In the Helmholtz flow, the velocity is constant along the streamlines bounding the dead air region.

Thus the surfaces of discontinuity leaving the leading and trailing edges may be represented by vortex sheets of uniform strength. In the development of the circulatory flow, on the other hand, a single surface of discontinuity or vortex surface is formed along the trailing edge as the wing starts from rest. For each increment in the strength of the circulation around the wing there appears a corresponding increase of the total vortex strength of this wake. However, as the flow approaches the steady state, the rate of growth of the circulation, and therefore the strength of the shed vortices, diminishes. In the final steady state of the two-dimensional flow no additional vortices leave the trailing edge.

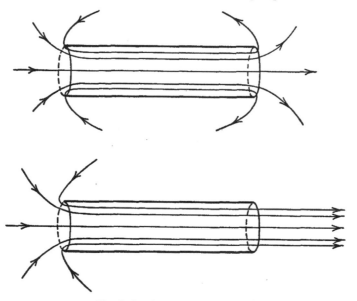

Fig. A,1b.   Flows through a pipe.

In the three-dimensional flow around a wing of finite span (Fig. A,1c) the vortex surface remains attached to the wing, but in the final steady state the discontinuity appears only in the lateral velocity v, so that the elementary vortices are left behind the wing in a direction parallel to the direction of motion.

Side edges.   The sharp distinction between the leading edge and the trailing edge in the Kutta-Joukowski flow cannot be carried over completely to actual wings of finite planform, especially if part of the outline of the wing lies parallel or nearly parallel to the stream direction. A satisfactory solution for the flow around such a side edge has not as yet been obtained. The calculations that have been made thus far are designed to satisfy the condition of zero lift at the side edge, but the solutions show infinite velocities *around* the edge (Fig. A,1d, top). A considera-

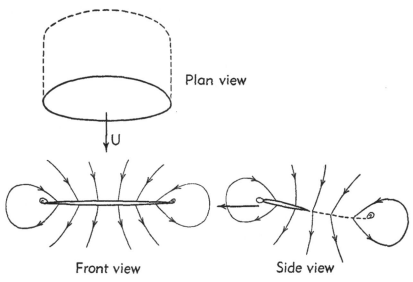

Plan view

Front view   Side view

Fig. A,1c.   Wake left by wing starting from rest.

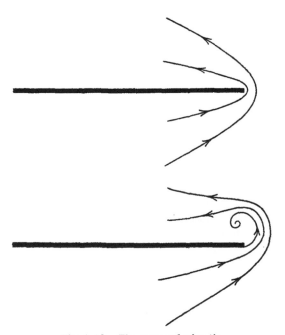

Fig. A,1d.   Flows around wing tip.

tion of the stresses involved, on the other hand, leads to the conclusion that the condition of tangential flow should, strictly speaking, be applied at the side edges as well as at the trailing edge (Fig. A,1d, bottom).

Experimental demonstration of airfoil efficiency. According to the Kutta-Joukowski hypothesis the two-dimensional frictionless flow around an airfoil results in the production of a lifting force without accompanying drag. The extent to which this ideal, dragless type of flow can be realized in practice under suitable conditions is well illustrated by an experiment performed by Jacobs [4] at the Langley Field Laboratory of the NACA. The airfoil (Fig. A,1e) in this experiment was of 5-foot chord, 12 per cent thickness, and extended completely across the wind tunnel. At an airspeed of 50 miles per hour and an angle of attack of 7°, a lift force of 100 pounds was measured. The drag amounted to only slightly more than one third of a pound—corresponding to a ratio of lift to drag of nearly 300 to 1. For comparative purposes a round rod having approximately the same drag in pounds as the lifting airfoil is also shown.

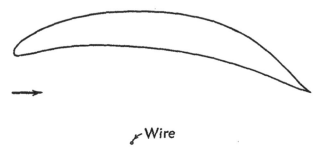

Fig. A,1e. Wing compared to circular wire of equal drag.

In modern wind tunnel experiments on more practical airfoils, a drag coefficient of 0.0045 and a lift-drag ratio of 100 to 1 is not unusual (see [5]). Such a drag coefficient, in the case of a 15 per cent thick airfoil, amounts to the force of the impact pressure acting on an area equal to only 3 per cent of the projected frontal area of the airfoil. Since this force is actually almost equal to the estimated surface friction, it is concluded that the type of pressure distribution obtained from the theory of perfect fluids can be very closely realized in practice. This of course is not true for bodies of a general or arbitrary shape, where because of flow separation the actual flow is frequently an exceedingly complex phenomenon, apparently not corresponding to any simple or intelligible solutions of the hydrodynamical equations.

ORIGIN OF DRAG IN THREE-DIMENSIONAL FLOW. In the experiments cited above, the conditions of two-dimensional flow were closely simulated by extending the wing completely across the wind tunnel. In free air the production of lift on a wing of finite span is accompanied by a pressure drag or "induced drag" which is associated with the continual extension of the vortex wake behind the wing discussed above. As a result of the induced drag, the lift-drag ratios of wings of practical proportions fall considerably below the values obtainable in two-dimensional flow.

Concept of induced drag. The flow around a lifting wing is characterized by an upwash in the stream ahead and a downwash behind. These upwash and downwash velocities may be represented by the action of a distribution of vortices in the airfoil; at a large distance the flow may be approximated by the velocity field of a single vortex superimposed on the velocity of the stream. It is easily seen that if the span of the wing extends to infinity the influence of the lifting vortex will be symmetrical ahead of and behind the center of lift. The absence of drag in two-dimensional lifting flow is related to this fore-and-aft symmetry of the downwash field.

If the span of the wing is limited, however, the fore-and-aft symmetry of the flow disappears. As explained by Lanchester [6], the vortices representing the circulatory flow around the wing must turn in the vicinity of the tips and extend backward, following the streamlines of the flow behind the wing. Turning the vortices from a lateral to a backward position diminishes the upwash in the field ahead of the wing and increases the downwash in the region behind. As a result of the downward inclination of the flow the lift of the section is reduced and inclined rearward, giving rise to a drag which is termed the "induced drag" [7].

Lifting line theory. It is seen that the reduction of upwash in the stream ahead of the wing is similar in effect to a reduction of the angle of attack. Prandtl [8] found that for wings of high aspect ratio a satisfactory representation of the flow and the drag could be obtained on this basis by applying a correction to the angle of attack in the two-dimensional formulas. The quantitative theory of this correction, known as the lifting line theory, is based on the assumption that the effect of the finite span is to give rise to a motion on a relatively large scale of dimensions, which can be superimposed on the smaller, locally two-dimensional wing section flows. For the purpose of computing this downwash the wing section flow is replaced temporarily by the flow around a single straight vortex at the center of pressure. Since a straight lifting vortex induces no downwash at its own axis, the downwash computed along the "lifting line" will arise solely from the vortices in the wake. At a great distance behind the wing, the wake forms a two-dimensional field of motion, and, since in the first order theory the form of the wake does not change with distance, it can be shown from symmetry that the value of the downwash at the wing is one half the final value in the two-dimensional field of the wake. Through the Helmholtz vortex theorem the distribution of vorticity in the wake is related to the gradient of the spanwise distribution of lift or circulation at the wing. Hence the induced downwash correction and the induced drag depend only on the spanwise distribution of the lift and are independent of the chordwise distribution of lift.

Although in its derivation the lifting line theory appears to involve drastic approximations, a comparison with more exact three-dimensional solutions, which will appear later in this section (Art. 7), indicates that in cases of straight narrow wings with smooth load distributions the theory is well founded. It may be shown from symmetry that the lifting line theory actually gives the correct value (in the sense of linearized lifting surface theory) of the downwash along the center of pressure of any distribution of lift that is symmetrical ahead of and behind a straight line at right angles to the direction of flight. However, for curved or yawed lifting lines and especially for wings of low aspect ratio, the lifting line theory is no longer adequate. As will be shown later in the present section the effect of compressibility at higher speeds is also such as to require a more complete treatment of the lifting surface in three dimensions.

A,2. Incompressible Flow in Two Dimensions. The Theory of Thin Airfoils.

Introduction. The necessity of minimizing the induced drag leads to the choice of a high aspect ratio for the proportions of practical wings designed for subsonic flight speeds. With such long narrow wings and especially at low speeds, the flow in the vicinity of the wing sections can be expected to resemble closely the flow around an infinitely long cylindrical wing of the same section profile. This fact has led to extensive use of the concept of two-dimensional flow in both theoretical and experimental studies of wing forms.

Mathematical formulation of two-dimensional flow fields. On the theoretical side, the assumption of two-dimensional flow simplifies the problem mathematically to the extent that the determination of the flow field and pressure distribution around any arbitrarily specified airfoil shape becomes a practical calculation. The mathematical simplification arises from the fact that the motion of an incompressible fluid in two dimensions can be represented by an analytic function of a single variable. Thus it can be shown by direct differentiation that the function

$$\varphi = F(\alpha x + \gamma z) \tag{2-1}$$

satisfies Laplace's equation

$$\frac{\partial^2 \varphi}{\partial x^2} + \frac{\partial^2 \varphi}{\partial z^2} = 0 \tag{2-2}$$

provided $\alpha^2 + \gamma^2 = 0$. The choice $\alpha = 1$, $\gamma = i$ leads to

$$\varphi = F(x + iz) \tag{2-3}$$

Every analytic function of such a complex variable can be separated into a pair of distinct functions, one real and one imaginary; thus

$$\varphi = \varphi_1(x, z) + i\varphi_2(x, z)$$

and both φ_1 and φ_2 will be solutions of Laplace's equation. Since φ_1 and φ_2 are orthogonal, they may represent the equipotential lines and the stream-lines of a two-dimensional flow.[1]

Linearization of boundary conditions. In the case of thin airfoils with small camber or angle of attack, a still further simplification of the two-dimensional potential flow theory is possible. In incompressible flow this simplification is the basis of the well-known Munk theory of thin airfoils [9], while at supersonic speeds it forms the basis of the Ackeret theory [10]. At the present time, engineering calculations for wings at high speeds are based almost exclusively on the assumptions of the thin airfoil theory, especially since, as will appear later in the present section, this theory offers a practical approach to the problem of compressible potential flow in three dimensions.

In its mathematical derivation the thin airfoil theory is shown to correspond to the limiting case of infinitesimal thickness or angle of attack. In this limiting case the boundary condition of the airfoil can be replaced by a specification of horizontal or vertical components of the perturbation velocity along a straight "chord line" of the airfoil, without regard for the actual displacement of the airfoil boundary away from this chord line. If the shape of the airfoil is given by the ordinates $z_1(x)$, then the requirement of tangential flow along the surface reduces to

$$\left(\frac{w}{U}\right)_{z=0} = \frac{dz_1}{dx} \tag{2-4}$$

The pressure distribution throughout the field is obtained from the horizontal perturbation velocity u by the relation

$$\frac{\Delta p}{\frac{1}{2}\rho_\infty U^2} = -\frac{2u}{U} \tag{2-5}$$

which is an approximate form of Bernoulli's equation.

The upper and lower surfaces of the chord line correspond to the upper and lower surfaces of the airfoil, and, since the velocities on these two surfaces are different, there will arise a discontinuity in the velocity across this line. A discontinuity in the vertical component w results from the thickness distribution of the airfoil, while a discontinuity in the horizontal component u arises from the camber or angle of attack of the mean line, and is related to the distribution of lift.

---

[1] The corresponding solution of Laplace's equation for three dimensions is

$$\varphi = F(\alpha x + \beta y + \gamma z)$$

with $\alpha^2 + \beta^2 + \gamma^2 = 0$. In this case, however, it is not generally possible to separate the solution into three parts corresponding to three distinct units α, β, γ.

With a fixed surface independent of the solution at which to consider the boundary velocities, and with a linear equation for the motion of the fluid (Laplace's equation, in this case), the airfoil problem becomes a linear one. Airfoil ordinates may be superimposed in any linear combination and the resulting velocity distribution may be obtained by the addition of the velocity fields corresponding to the elements of the combination. This simplification makes it possible to divide airfoil problems into two classes, i.e. those involving only camber or angle of attack of the mean line (considered to be an infinitely thin airfoil), and those corresponding to distributions of thickness disposed symmetrically above and below the chord line. In this way the discontinuities in u and w can be treated separately. In the former case the horizontal perturbation velocity u will have equal and opposite values above and below the chord line, resulting in a lifting pressure given in coefficient form by

$$C_p = \frac{\Delta p_{\text{lower}} - \Delta p_{\text{upper}}}{\frac{1}{2}\rho_\infty U^2} = \frac{4u}{U} \tag{2-6}$$

where u is taken on the upper side. With a symmetrical distribution of thickness, the horizontal perturbation velocity u will be continuous, giving rise to equal values of the pressure above and below the chord plane and hence to no lift. The vertical velocity w is discontinuous and is related to the thickness by Eq. 2-4.

Complex velocity functions.   As noted above, any analytic function of a complex variable can represent the streamlines and equipotential lines of a two-dimensional flow field. Differentiation for the velocity components shows that the conjugate of the complex velocity vector can be given directly by such a function; i.e.

$$\frac{u - iw}{U} = f(x + iz) \tag{2-7}$$

Thin airfoil flows can hence be obtained by constructing certain functions f which vanish at infinity and which show the desired distribution of velocities along the chord line. For convenience in writing such functions the chord line is taken along the real axis between ± 1.

As an example, suppose it is desired to obtain the solution for a flat surface. Since $w/U = dz/dx = $ const, it is necessary to find a function whose imaginary part remains at a fixed value along the real axis between ± 1. One such function[2] is

$$\frac{u - iw}{U} = \frac{1}{\sqrt{1 - \xi^2}} \tag{2-8}$$

---

[2] It should be noted that the problem is not completely specified and that other flows about a flat surface are possible.

where $\xi = x + iz$. The real part is discontinuous; i.e.,

$$u(x + 0 \cdot i) = -u(x - 0 \cdot i)$$

between the points ± 1, while the imaginary part is continuous all along the real axis. Since the imaginary part is zero between ± 1, the solution represents a lifting flow over a flat plate at zero angle of attack and is, in fact, the well-known circulatory flow around the flat plate. The total circulation is given by

$$\Gamma = \int_{-1}^{+1} 2u(x + 0 \cdot i)dx = 2\pi U \tag{2-9}$$

in this case.

The following table lists a number of complex velocity functions for thin airfoils[3] and shows the corresponding distributions of horizontal and vertical perturbation velocities along the chord line.

Application to airfoils with finite thickness. The interpretation of thin airfoil flows in terms of airfoils of finite thickness can be facilitated by a

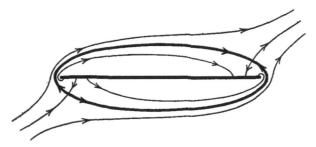

Fig. A,2a. Internal streamlines.

comparison of these flows with the known exact solutions for typical shapes. Most simple shapes such as the Joukowski airfoil and the ellipse can be produced by the action of a continuous distribution of sources and vortices along a mean line inside the body. The internal flows in such cases therefore show a discontinuity of velocity along this line. When comparison is made with the velocity distribution given in each case by thin airfoil theory, a surprisingly close similarity is found to exist. The velocity distributions given by the thin airfoil theory along the line $z = 0$ may thus be considered as the velocities of the internal flow along the mean line of an airfoil of finite thickness. (See Fig. A,2a.)

In the case of the elliptical section, it is found that the velocity function actually satisfies the boundary condition for an elliptical shape of slightly different proportions than those given by the approximate

---

[3] Most of the functions listed may be derived from formulas given in *NACA Rept. 833* by H. J. Allen, *Fundamentals of Fluid Dynamics for Aircraft Designers* by M. M. Munk, or *The Elements of Aerofoil and Airscrew Theory* by H. Glauert. (See Bibliography, Art. 16.)

Table A,2. Two-dimensional flow functions.

Velocity function $f(\xi) = \dfrac{u - iw}{U}$	Components u/U, w/U along chord line	Airfoil shape $z = \int \dfrac{w}{U}\, dz$
1. $\dfrac{i}{\sqrt{\xi^2 - 1}} = \dfrac{-1}{\sqrt{1 - \xi^2}}$		
2. $i\left(1 - \dfrac{\xi}{\sqrt{\xi^2 - 1}}\right)$		

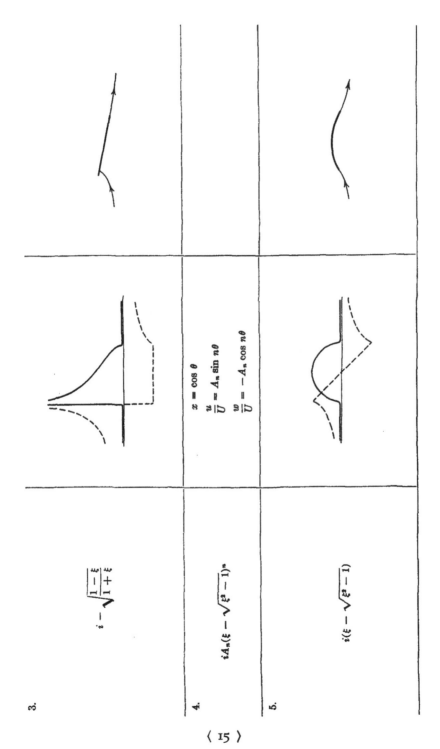

3. $i - \sqrt{\dfrac{1-\xi}{1+\xi}}$

$x = \cos\theta$

$\dfrac{u}{U} = A_n \sin n\theta$

$\dfrac{w}{U} = -A_n \cos n\theta$

4. $iA_n(\xi - \sqrt{\xi^2 - 1})^n$

5. $i(\xi - \sqrt{\xi^2 - 1})$

Velocity function $f(\xi) = \dfrac{u - iw}{U}$	Components u/U, w/U along chord line	Airfoil shape $z = \int \dfrac{w}{U}\, dx$
6. $-i(\xi - \sqrt{\xi^2 - 1})^2$		
7. $\dfrac{2i}{\pi} A_n Q_n(\xi)$	$\dfrac{u}{U} = A_n P_n(x)$ $\dfrac{w}{U} = \dfrac{2}{\pi} A_n Q_n(x)$	
8. $\dfrac{2i}{\pi} Q_0(\xi)$		

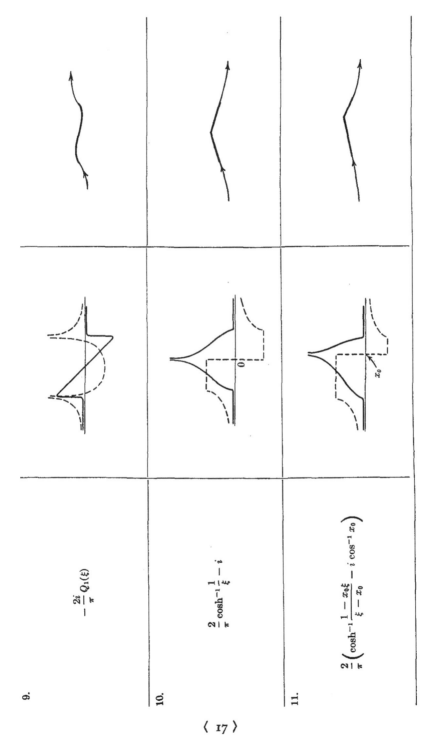

9.

$$-\frac{2i}{\pi} Q_1(\xi)$$

10.

$$\frac{2}{\pi} \cosh^{-1} \frac{1}{\xi} - i$$

11.

$$\frac{2}{\pi} \left(\cosh^{-1} \frac{1 - x_0 \xi}{\xi - x_0} - i \cos^{-1} x_0 \right)$$

	Velocity function $f(\xi) = \dfrac{u - iw}{U}$	Components u/U, w/U along chord line	Airfoil shape $z = \displaystyle\int \dfrac{w}{U}\, dx$
12.	$\dfrac{1}{\sqrt{\xi^2 - 1}} = \dfrac{i}{\sqrt{1 - \xi^2}}$		
13.	$1 - \dfrac{\xi}{\sqrt{\xi^2 - 1}}$		

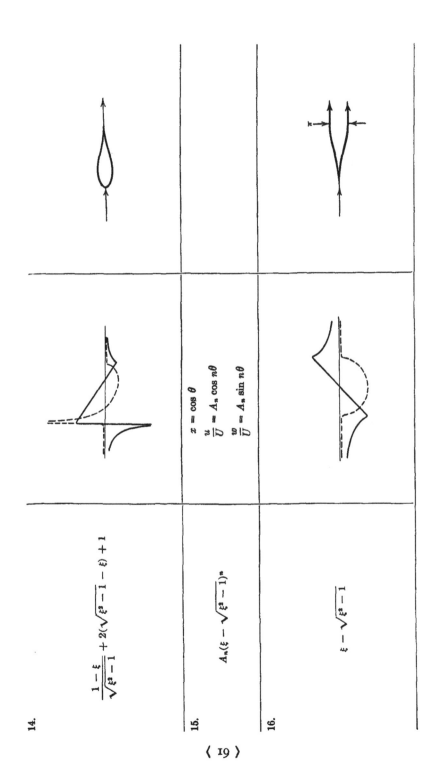

14.

$$\frac{1-\xi}{\sqrt{\xi^2-1}}+2(\sqrt{\xi^2-1}-\xi)+1$$

15.

$$A_n(\xi-\sqrt{\xi^2-1})^n$$

$$x=\cos\theta$$
$$\frac{u}{U}=A_n\cos n\theta$$
$$\frac{w}{U}=A_n\sin n\theta$$

16.

$$\xi-\sqrt{\xi^2-1}$$

Velocity function $f(\xi) = \dfrac{u - iw}{U}$	Components u/U, w/U along chord line	Airfoil shape $z = \int \dfrac{w}{U}\, dx$
17. $-(\xi - \sqrt{\xi^2 - 1})^2$		
18. $\dfrac{2}{\pi} A_n Q_n(\xi)$	$\dfrac{u}{U} = \dfrac{2}{\pi} A_n Q_n(x)$ $\dfrac{w}{U} = A_n P_n(x)$	
19. $\dfrac{2}{\pi} Q_0(\xi)$		

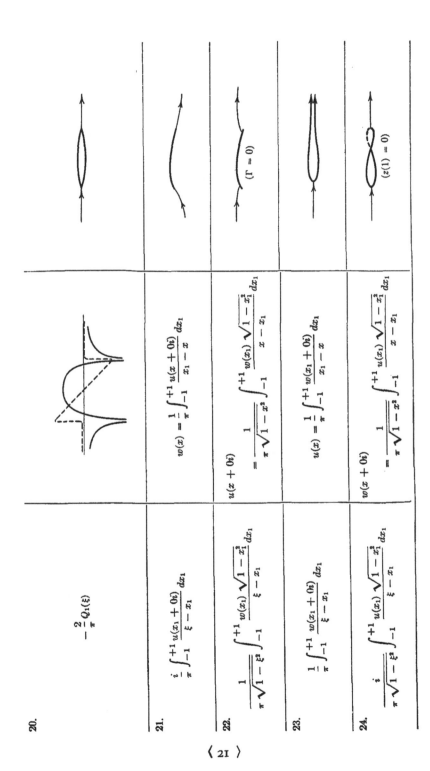

20.

$$-\frac{2}{\pi} Q_1(\xi)$$

21.

$$w(x) = \frac{1}{\pi} \int_{-1}^{+1} \frac{u(x + 0i)}{x_1 - x} dx_1$$

$$\frac{i}{\pi} \int_{-1}^{+1} \frac{u(x_1 + 0i)}{\xi - x_1} dx_1$$

22. $(\Gamma = 0)$

$$u(x + 0i) = \frac{1}{\pi \sqrt{1 - x^2}} \int_{-1}^{+1} w(x_1) \frac{\sqrt{1 - x_1^2}}{x - x_1} dx_1$$

$$\frac{1}{\pi \sqrt{1 - \xi^2}} \int_{-1}^{+1} w(x_1) \frac{\sqrt{1 - x_1^2}}{\xi - x_1} dx_1$$

23.

$$u(x) = \frac{1}{\pi} \int_{-1}^{+1} \frac{w(x_1 + 0i)}{x_1 - x} dx_1$$

$$\frac{1}{i\pi} \int_{-1}^{+1} \frac{w(x_1 + 0i)}{\xi - x_1} dx_1$$

24. $z(1)z(0) = 0$

$$w(x + 0i) = \frac{1}{\pi \sqrt{1 - x^2}} \int_{-1}^{+1} u(x_1) \frac{\sqrt{1 - x_1^2}}{x - x_1} dx_1$$

$$\frac{i}{\pi \sqrt{1 - \xi^2}} \int_{-1}^{+1} u(x_1) \frac{\sqrt{1 - x_1^2}}{\xi - x_1} dx_1$$

boundary condition. The boundary of the actual ellipse extends beyond the ends of the chord line, so that the points ± 1 occur at the foci of the elliptical section. By substituting the actual coordinates of the elliptical boundary into formula 13 of Table A,2, the velocity distribution around the surface may be obtained. This same function can be made to satisfy the boundary condition for every confocal ellipse by adjusting the magnitude of the velocity discontinuity, or source strength, in relation to the stream velocity U.

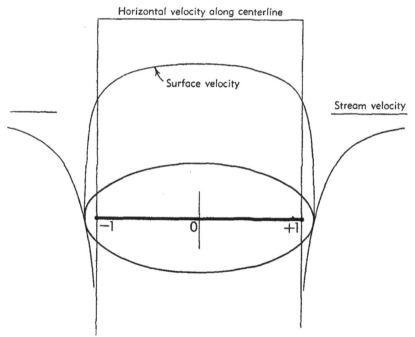

Fig. A,2b.   Comparison of velocity distributions along the chord line and at the surface of an elliptical section.

Fig. A,2b illustrates the nature of the correspondence between the velocities along the axis and the surface velocities in the case of a thick section. The effect of proceeding from the chord line to the surface may be described as an attenuation with distance. For thin sections the attenuation is small, but is always more pronounced in regions of high local velocities or accelerations. Thus the infinite velocities at the ends of the chord line, ± 1, are reduced to finite values at the position of the nose. The fact that such infinite velocities arise only in the fictitious "internal" flow removes a difficulty in the physical interpretation of the theory and is especially important in the application of the theory to compressible fluids.

Total lift and drag. Leading edge forces. The total lift may be obtained, to the first order, by an integration of the pressure distribution along the chord line, and the total drag by integrating the product of the pressure and the local inclination of the airfoil surface. In the integration for the drag, however, care must be taken to evaluate the effects of singularities which appear—usually at the leading edge—in the thin airfoil flows. Such singularities can be included by integrating along a path $\xi = x + 0 \cdot i$ displaced a vanishingly small distance in the z direction and also slightly extended so as to pass above the point of infinite velocity at the leading edge. Since $\Delta p = -\rho_\infty u U$ and $dz/dx = w/U$, the integral for the drag may be written, considering both upper and lower surfaces,

$$D = 2 \int \Delta p \frac{dz}{dx} \, dx = -2 \int \rho_\infty u w \, dx \qquad (2\text{-}10)$$

For the two-dimensional velocity field, the product occurring in the integral on the right may be denoted by

$$-2uw = \text{I.P.} \ (u - iw)^2 \qquad (2\text{-}11)$$

In the case of the flat plate at an angle of attack α, the integral for the drag becomes (see formula 3, Table A,2)

$$D = \text{I.P.} \int \rho_\infty U^2 \alpha^2 \left(\frac{1 - \xi}{1 + \xi} - 2i \sqrt{\frac{1 - \xi}{1 + \xi}} - 1 \right) d\xi \qquad (2\text{-}12)$$

The leading edge force is obtained by evaluating the contribution to this integral in passing just above the singular point at $\xi = -1$. In the limit, only the first term in the parentheses is effective, and its integral approaches the value $2 \ln (1 + \xi)$. In crossing over the point -1, the vector $(1 + \xi)$ rotates through $-180°$, so that the imaginary part of the logarithm changes by $-\pi$. Hence the contribution of the leading edge singularity to the drag is

$$D_{1.e.} = -2\pi \rho_\infty U^2 \alpha^2 \qquad (2\text{-}13)$$

Since

$$L = 2\pi \rho_\infty U^2 \alpha \qquad (\text{chord} = 2)$$

the leading edge force is equal and opposite to the product of the total lift and the slope of the wing surface, and the total drag in this case is zero.

The foregoing development can be generalized as follows: Whenever the pressure in the vicinity of the leading edge ($x = -1$) can be written in the form

$$-\rho_\infty U \left[\frac{f(x)}{\sqrt{1 + x}} + \text{nonsingular terms} \right]; \quad f(-1) \text{ finite}$$

or, from Eq. 2-7, the complex velocity function approaches, in the neighborhood of the leading edge,

$$u - iw = \frac{f(\xi)}{\sqrt{1 + \xi}} + \text{nonsingular terms}$$

the leading edge force will be

$$D_{\text{l.e.}} = \text{I.P. } \rho_\infty \int \left[\frac{f^2(\xi)}{1 + \xi} + \text{nonsingular terms} \right] d\xi$$

In the limit this gives

$$D_{\text{l.e.}} = -\pi \rho_\infty C^2 \tag{2-14}$$

in which C is the value, at the leading edge, of u times the square root of the distance to the leading edge.

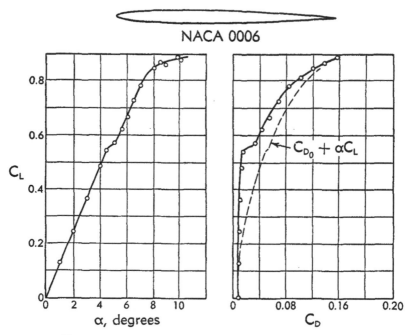

Fig. A,2c. Experimental lift and drag curves for thin section.

A similar limiting process must be considered in connection with the velocity functions associated with distributions of thickness with rounded leading edges, such as formula 14, Table A,2. In these cases, Eq. 2-14 will be found to give a positive component of drag.

With an airfoil of finite thickness and a continuously curved surface, the leading edge force cannot of course be clearly separated from the force developed on other parts of the airfoil surface. If the airfoil is made thinner, however, the suction developed in the upward flow around the leading edge becomes greater, while the forwardly projecting area over which this suction can act becomes less. The two factors change in such a way that their product remains essentially constant right down to the limit of zero thickness. In the limiting case of zero thickness the pressure stress in the vicinity of the nose tends toward infinity.

Such a limit cannot of course be approached in practice. It is found experimentally that if the radius of the nose is made too small, flow separation will occur at a relatively small angle of attack. Increases of angle of attack beyond this point cause a rapid increase in drag coefficient.

Physical realization of foregoing type of flow. Fig. A,2c, taken from experimental data given in [11], illustrates the rapid rise in drag encountered on an airfoil of 6 per cent maximum thickness when the flow ceases to follow the sharp curvature around the nose. It will be noted that the drag remains nearly constant up to a lift coefficient of 0.55 (angle of attack of 5°). This behavior illustrates the action of the leading edge suction force. A small increase of the angle of attack beyond this point, however, causes flow separation at the nose, and the resultant force on the airfoil falls back toward a perpendicular to the chord direction, as illustrated by the rapid increase of the drag.

In general, the range of angles of attack and thickness-chord ratios within which an unseparated flow, resembling the potential flow pattern, can be maintained constitutes the range of practical flight conditions. Outside this range the drag becomes excessive, while the control and flying qualities are usually impaired by unsteadiness of the flow and buffetting.

A,3. Effects of Compressibility at Subsonic Speeds.

INTRODUCTION. As long as the speed of flight is small compared with the velocity of sound and the pressures generated are small relative to the ambient atmospheric pressure, the influence of the compressibility of the air can safely be neglected. The development of power plants of increased capacity, however, continually makes possible flight at higher altitudes and higher speeds. It is found that, up to a certain limit, the higher speeds can be achieved without loss of efficiency or without increase of the amount of fuel expended per mile of flight. Beyond this limit, however, the favorable aerodynamic properties of conventional wing forms, and hence the high ratio of lift to drag of the airplane, deteriorate rapidly. Since the limitation is imposed by the effects of compressibility, the study of such effects has become a primary task of present-day aerodynamic research.

PHYSICAL CONSIDERATIONS. Before proceeding to a quantitative study of the effects of the compressibility of air on the behavior of an airfoil, it is desirable to review certain physical aspects of the flow of an incompressible fluid. Suppose the airfoil, as in Fig. A,3a, to be immersed in a uniform stream moving to the right. Since its thickness has the effect of diminishing the cross section of the oncoming stream, the flow around the midsection of the airfoil occurs with an excess of velocity and in this region the mass flux per unit area must be increased to prevent the accumulation of mass in the region ahead. Assuming a frictionless fluid, the motion of each element of volume must be produced by the action of pressure, or buoyancy, forces. Thus the flow is accelerated past the midsection of the airfoil by entering a region of reduced pressures. The areas of positive and negative pressure are symmetrically disposed, so as to conduct the flow around the airfoil without the occurrence of any pressure drag. It is also readily apparent from geometric considerations that the widening of the stream tubes near the nose and the contraction of width in the regions of increased velocity above and below the airfoil lead to a progressive reduction in curvature of the streamlines, and hence to an attenuation of the flow disturbance with distance from the airfoil.

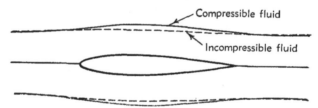

Fig. A,3a. Expansion of streamlines in compressible flow.

In the corresponding flow of an elastic fluid, the streamline pattern is modified by the expansion of the volume of the fluid on passing through the regions of low pressure above and below the airfoil and, to a lesser degree, by the compression occurring in regions of higher pressure. Since its volume is increased in the regions of high velocity and low pressure, the elastic fluid requires a still higher velocity in order to get past the midsection of the airfoil, as well as more space in order to maintain the given total mass flow. The expansion of the minimum cross section of the stream tubes forces the streamlines outward to conform more nearly to the curvature of the airfoil surface, with the result that the disturbance caused by the airfoil extends vertically to a greater distance.

ONE-DIMENSIONAL ANALYSIS. In spite of its fairly simple physical aspect, the problem of determining the configuration of the flow field around an airfoil so as to agree with the dynamical equations and with a realistic equation of state for air has turned out to be one of great mathematical difficulty. A quantitative insight into the phenomenon can be gained, however, through the so-called "one-dimensional" analysis [12]. This analysis is based on the selection of stream tubes of infinitesimal cross section, so that the change in any quantity across the width of the stream tube is negligible. The velocity, pressure, and density can then be considered as functions of the distance along the stream tube. Such one-dimensional relations must be satisfied throughout any steady flow, and hence form a necessary condition on the field as a whole—but they are not sufficient to determine the complete pattern because of the additional requirement of dynamical balance between adjacent stream tubes.

Assuming frictionless motion, we have for the dynamical equation

$$u \frac{du}{dx} = -\frac{1}{\rho} \frac{dp}{dx} \tag{3-1}$$

where the velocity u and the distance x are measured along the stream tube. If A is the cross-sectional area of the stream tube, the equation for continuity of mass flow is

$$\frac{d}{dx}(\rho A u) = 0 \tag{3-2}$$

By making use of the expression for the velocity of sound in an elastic medium, i.e.

$$a = \sqrt{\frac{\partial p}{\partial \rho}} \tag{3-3}$$

Eq. 3-1 and 3-2 can be combined into a single relation between the velocity of flow, the rate of change of cross section, and the local Mach number u/a, viz.

$$\left(1 - \frac{u^2}{a^2}\right) \frac{du}{dx} = -\frac{u}{a} \frac{dA}{dx} \tag{3-4}$$

Eq. 3-4 shows that the stream tubes become narrower as the velocity increases, so long as the ratio u/a remains less than one. Just at the speed of sound ($u = a$) the quantity dA/dx becomes zero, and at this speed a decrease of pressure in the direction of flow causes an expansion of volume just sufficient to counteract the increased flow velocity, so that the streamlines are no longer reduced in width by an increase of velocity. When the local velocity reaches sonic speed the rate of mass flow per unit of cross-sectional area ρu reaches an absolute maximum and can no longer be increased by the action of a pressure gradient. At this speed any change of velocity through the action of pressure forces tends to diminish the mass flowing through a given cross section.

At the exact speed of sound the idealization of a steady flow which vanishes at infinity breaks down, since mass would continually accumulate ahead of the obstacle. This accumulation makes itself apparent in the phenomenon of "choking" in the wind tunnel and in the extensive bow wave developed ahead of an airplane when flying near the speed of sound [13].

LINEARIZED COMPRESSIBLE FLOW THEORY.

Similarity of compressible and incompressible flows at subsonic speeds. Within the regime of continuous subsonic motion, it is found that the flow of a compressible fluid retains a basic similarity to an incompressible flow. In particular, the compressible flow can for many purposes be considered an irrotational potential motion. In addition to the absence of friction, the existence of potential motion in a compressible fluid depends on the existence of a unique relation between the pressure and the density. In a frictionless flow the fluid elements are accelerated entirely by the action of the pressure gradient. If the density is a function of the pressure alone, the pressure gradient will coincide in direction with the density gradient at all points. The force on each element will then be aligned with the center of gravity of the element and the pressure forces will introduce no rotation.

Prandtl's linerarized equation for steady flow. Prandtl has shown [14] that the assumptions of the thin airfoil theory lead to an essential simplification of the equation for the steady potential flow of a compressible fluid. Since successful flight at high speeds actually depends on the use of thin wings and slender bodies, the resulting linearized theory, while not suitable for the study of compressible flows in general, appears to provide a suitable basis for the practical calculations of aeronautical engineering.

It is instructive to derive Prandtl's approximate linearized equation of motion from the more exact one-dimensional flow equation (3-4). It will be evident that the rate of change of the cross-sectional area A of a stream tube can be related kinematically to small lateral and vertical velocities v and w superimposed on the main velocity u. By such a kinematic relation the quantity $(u/A)(\partial A/\partial x)$ is found to be equal to $\partial v/\partial y + \partial w/\partial z$, or, in other words, to the divergence of the velocity pattern observed by looking along the local direction of the streamlines. Hence Eq. 3-4 can be written

$$\left(1 - \frac{u^2}{a^2}\right)\frac{\partial u}{\partial x} + \frac{\partial v}{\partial y} + \frac{\partial w}{\partial z} = 0 \qquad (3\text{-}5)$$

Eq. 3-5 must be satisfied locally at every point in the flow field and, in order to apply it to the flow field as a whole, variations of the local Mach number u/a as well as the inclination of the axes x, y, z (which have been assumed to lie parallel to the local flow direction) would have to be considered. However, if the airfoil is thin and its angle of attack small, so that the streamlines are nearly horizontal, little error is made in the differential lengths if a single set of x, y, z axes aligned with the main stream velocity U is used for the entire flow field. Furthermore, if the resultant velocity throughout the field departs only slightly from the free stream velocity, the local Mach number u/a will depart only slightly from the free stream Mach number M_∞.

In addition to the one-dimensional equation for the flow along the streamlines, it is necessary to introduce the condition of dynamical balance between adjacent streamlines. Thus we must have

$$\rho_\infty U \frac{\partial w}{\partial x} = -\frac{\partial p}{\partial z} \tag{3-6}$$

in order for the pressure gradients to balance the centrifugal forces. As in the case of incompressible frictionless flow, it is found that to the first order $\Delta p = -\rho_\infty u U$, so that Eq. 3-6 becomes

$$\frac{\partial w}{\partial x} = \frac{\partial u}{\partial z} \tag{3-7}$$

The latter equation expresses the condition of irrotationality and leads to the existence of a velocity potential ϕ. Introducing $\partial u/\partial x = \phi_{xx} = \varphi_{xx}$, etc. into Eq. 3-5 results in

$$(1 - M_\infty^2)\varphi_{xx} + \varphi_{yy} + \varphi_{zz} = 0 \tag{3-8}$$

which is the equation given by Prandtl.

Relation to acoustic equation. Eq. 3-8 may be shown to be a special case of the equation for sound waves of small amplitude, i.e.

$$\varphi_{xx} + \varphi_{yy} + \varphi_{zz} = \frac{1}{a_\infty^2} \varphi_{tt} \tag{3-9}$$

From this point of view the moving airfoil is considered to give rise to small velocities and pressures in air that is otherwise at rest. If the airfoil has been moving at constant velocity for a sufficiently long time it may be assumed to have developed a fixed, or steady flow pattern, which travels with the airfoil at the velocity of flight U. By adopting axes fixed in the moving airfoil, so that $x' = x + Ut$, φ_{xx} may be replaced by $\varphi_{x'x'}$, and φ_{tt} by $U^2\varphi_{x'x'}$. Then, transposing the term on the right, replacing U^2/a_∞^2 by M_∞^2, and dropping the primes reduces Eq. 3-9 to Eq. 3-8. The steady flow around the thin airfoil may thus be considered as a type of sound disturbance that travels at a velocity U without change of form.

It is of interest to note the connection between the one-dimensional flow equation and the theory of acoustic waves. Such waves present an infinite variety of motions, but at the wave front, which travels at the acoustic velocity a, all lateral or vertical perturbation velocities disappear and the disturbance consists solely of longitudinal motions. Hence a sound wave is commonly referred to as a "longitudinal" wave. For that part of the wave which travels at the acoustic velocity a the streamlines are thus straight and parallel, as required by the one-dimensional equation.

Correspondence between compressible and incompressible flows. Eq. 3-8 can be reduced to Laplace's equation by a simple transformation of coordinates amounting to a change in the ratio of the horizontal to the vertical and lateral dimensions of the potential field. This may be seen by rewriting Eq. 3-8 in the form

$$\varphi_{xx} = -\frac{(\varphi_{yy} + \varphi_{zz})}{1 - M_\infty^2} \tag{3-10}$$

Comparison with Laplace's equation

$$\varphi_{xx} = -(\varphi_{yy} + \varphi_{zz}) \tag{3-11}$$

shows that in the compressible flow the second derivative of φ in the x direction is increased by the factor $1/(1 - M_\infty^2)$. This is just the effect

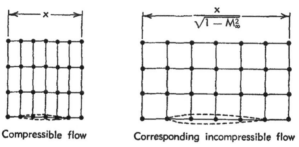

Compressible flow   Corresponding incompressible flow

Fig. A,3b. Distribution of points having equal values of ϕ in a Prandtl transformation.

that would be produced by taking a distribution of φ and compressing it uniformly in the x-wise direction by the factor $\sqrt{1 - M_\infty^2}$ (see Fig. A,3b). Thus, if a solution of Laplace's equation for incompressible flow is known, a corresponding solution for the linearized compressible flow equation can immediately be produced.

If the potential distribution satisfying Laplace's equation and the corresponding foreshortened distribution (which satisfies the compressible flow equation) are compared at points having the same values of φ, it is seen that $u = \partial\varphi/\partial x$ is increased in the ratio $1/\sqrt{1 - M_\infty^2}$, while the lateral and vertical velocities, corresponding to the lateral and vertical gradients of φ, remain unchanged.

Although the construction of a flow field for a compressible medium from a known incompressible flow is relatively simple, care must be taken in the determination of the boundary conditions satisfied by the compressible flow field. Suppose now a solution of Laplace's equation satisfying the boundary condition for a certain thin airfoil is given. If the airfoil chord line and the surrounding potential field are foreshortened together, the values of $w = \partial\varphi/\partial z$ will remain unchanged at corresponding points along the chord line. Considering the same value of U in the two cases, it follows that the slopes of the streamlines emanating from the chord plane will be the same at equal fractional distances along the chord in the two cases. Within the usual approximation of the thin airfoil theory this slope is equal to the slope of the airfoil surface and the two airfoil sections are thus geometrically similar. It is clear that in three-dimensional flow a change in aspect ratio will be involved. In two-dimensional flow, however, the foreshortening of the chord corresponds merely to a change in the absolute size of the wing, so that the effect of compressibility on an airfoil section of given slope is simply to increase the horizontal perturbation velocities over the airfoil surface uniformly by the factor $1/\sqrt{1 - M_\infty^2}$. This factor was also found by Glauert [15] and is frequently referred to as the Prandtl-Glauert correction.

More Accurate Representations of Subsonic Compressible Flow. Although the Prandtl-Glauert method provides an exceedingly simple means for calculating the flow around an airfoil in two dimensions, it should be borne in mind that this method generally underestimates the effect of compressibility on the magnitude of the disturbance with airfoils of finite thickness. As the stream Mach number approaches 1, the quantity $1/\sqrt{1 - M_\infty^2}$ approaches infinity, with the result that the ratio of the horizontal perturbation velocity u to the airfoil thickness also approaches infinity. Obviously for any finite thickness a point will be reached at which the perturbation velocity u cannot be considered small relative to the stream velocity. Hence the Prandtl-Glauert formula begins to show increasing departures from reality at some Mach number below unity, the exact value of which depends on the thickness or angle of attack of the airfoil.

More accurate representations of the two-dimensional flow around airfoils of finite thickness at subsonic Mach numbers have been the subject of extensive investigation (see VI,E and F). In these investigations the assumption of isentropic potential motion is retained and solutions of the steady flow equation [14]

$$\left(1 - \frac{u^2}{a^2}\right)\frac{\partial u}{\partial x} + \left(1 - \frac{w^2}{a^2}\right)\frac{\partial w}{\partial z} - \frac{2uw}{a^2}\frac{\partial u}{\partial z} = 0 \qquad (3\text{-}12)$$

together with the irrotationality relation

$$\frac{\partial u}{\partial z} = \frac{\partial w}{\partial x} \qquad (3\text{-}13)$$

are sought. In these equations u and w refer, of course, to the components of the resultant velocity rather than to small perturbation velocities, and the velocity of sound a is a function of the local velocity at each point.

Exact solutions of Eq. 3-12 and 3-13 are supplied by the hodograph method of Molenbroeck [16] and Chaplygin [17]. Although the method cannot be carried out exactly for an airfoil of fixed shape, its application to the airfoil problem has been studied by von Kármán and Tsien [18], and others (see VI,F,5). The results of these investigations show that the compressible flow over an airfoil section for a wide range of Mach numbers can be obtained, at least approximately, by a simple correction applied to the velocities and pressures of the incompressible flow over the same shape. In terms of the pressure coefficient $C_p (= \Delta p / \frac{1}{2}\rho_\infty U^2)$, von Kármán and Tsien give the correction

$$C_p = \frac{C_{p_0}}{\sqrt{1 - M_\infty^2} + \dfrac{C_{p_0}}{2}(1 - \sqrt{1 - M_\infty^2})} \tag{3-14}$$

where M_∞ is the free stream Mach number and C_{p_0} is the pressure coefficient at $M_\infty = 0$. It appears that the correction is greater in regions of higher local velocity, a result in agreement with the one-dimensional flow equation (3-4). Experiments show that the Kármán-Tsien correction affords a noticeable improvement over the linearized theory for airfoils of finite thickness in two-dimensional flow, but the method likewise fails after a certain critical Mach number is exceeded.

A method [19] adaptable to sufficiently thin shapes involves the direct solution of Eq. 3-12 by an expansion of the velocity potential in powers of the thickness or angle of attack. This method amounts to a successive improvement of the linear approximation, which corresponds to the *first* power of the thickness-chord ratio t/c. The result obtained by Kaplan for the lift curve slope of elliptical sections at $\alpha = 0$ may be written

$$\frac{L_c}{L_i} = \mu + \frac{t/c}{1 + (t/c)}\left[\mu(\mu + 1) + \frac{\gamma + 1}{4}(\mu^2 - 1)^2\right] \tag{3-15}$$

where $\mu = 1/\sqrt{1 - M_\infty^2}$ and γ is the ratio of specific heats. A similar expression was found by Hantzsche and Wendt [20] for Joukowski sections.

If the assumptions of infinitesmal perturbation velocities and two-dimensional flow are retained and the stream velocity is allowed to approach sonic speed, it is found that the zone of disturbance expands vertically without limit and that the ratio of the longitudinal to the vertical perturbation velocities increases indefinitely. A simplification of the differential equation which makes use of this change in the geometric and kinematic proportions of the flow field has recently been studied by von Kármán [21], Guderley [22], and Oswatitsch [23, 24]. If the velocity field is supposed to be nearly horizontal and to consist of small increments above and below the critical velocity of sound, an especially simple nonlinear differential equation is obtained. This equation, known as

Tricomi's equation, seems to express the essential features of a transonic flow field about thin sections in two-dimensional flow and, at the present time, investigations of the practical validity of the method are under way (VI,A,7 and VI,F,11).

LIMITATIONS OF PRESENT METHODS OF ANALYSIS. Thus far, theoretical investigations of flows at high subsonic velocities have been limited for the most part to cases of two-dimensional motion. As the critical Mach number is approached, it is found that the region of reduced density and increased velocity expands vertically at a rapid rate, so that the size of the disturbed region grows out of proportion to the size of the body and the relation of the pattern to the shape of the body tends to become somewhat indefinite. As noted above, the field begins to resemble that of a longitudinal sound wave which, once started, travels without the assistance of a moving solid. In part, the difficulty of the solution must be attributed to the assumed two-dimensional character of the motion, since this assumption effectively limits the expansion of the flow to the vertical dimension. It can be expected that in three dimensions the critical flow condition will be delayed to a higher Mach number and this is indeed found to be the case experimentally [25]. Studies of simplified flow equations for partially supersonic flows in three dimensions have recently been made by Spreiter [26] and Berndt [27].

When solutions obtained on the basis of steady potential flow are compared with observed flow patterns over sections of normal thickness, a fundamental discrepancy is encountered. Experiments in two-dimensional flow, or with straight wings of high aspect ratio, show that the continuous regular flow pattern around the airfoil changes rather abruptly to a discontinuous and frequently unsteady motion shortly after the appearance of a zone of supersonic velocity in the flow field. The supersonic zone expands rapidly and is followed by a vertically spreading shock wave. The fore-and-aft symmetry associated with the incompressible, or purely subsonic, type of flow is then lost, and a rapid increase of the drag coefficient appears. Solutions of the differential equations (3-12 and 3-13), however, continue to show reversible flow patterns and an absence of drag for flows containing extensive regions of supersonic velocity. It has not been possible to find experimentally the conditions under which this dragless type of flow might be realized in practice. (See VI,F,13.)

As far as can be determined from the differential equation (3-12), the condition of fore-and-aft symmetry would also exist in the flow over a symmetrical body in supersonic flight. Thus the solution would show Mach waves extending forward from the body as well as to the rear, and would be obviously incorrect from the physical standpoint. No such definite criterion for discarding the symmetrical solution in the subsonic case has as yet been established, although recent investigators [28,29,30]

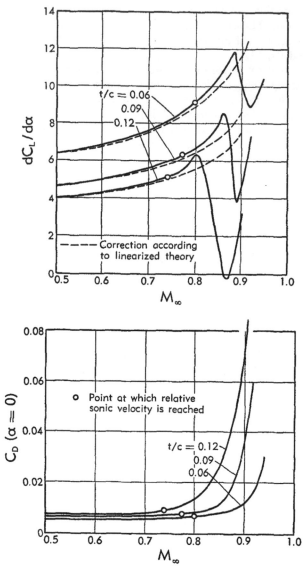

Fig. A,3c. Effect of airfoil thickness on lift curve slope
and drag at subsonic Mach numbers [31].

have indicated that this type of configuration may be unstable if small disturbances in the flow or in the shape of the boundary are considered (VI,F,14 and 15).

INDICATION OF EXPERIMENT. Wind tunnel measurements on wings of conventional form and thickness show a large increase of the drag coefficient and erratic variations of the lift and center of pressure after the development of shock waves, and schlieren photographs show that the shock wave usually causes flow separation.

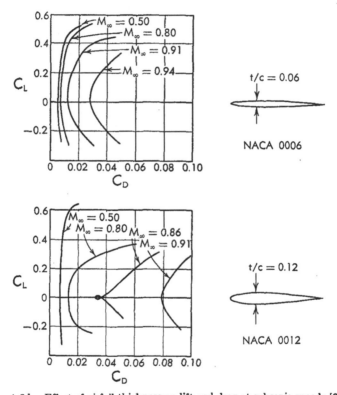

Fig. A,3d.   Effect of airfoil thickness on lift and drag at subsonic speeds [*31*].

Fig. A,3c and A,3d show typical values of the lift and drag of conventional wings in this speed range. It will be noted that the deterioration of the aerodynamic qualities does not begin immediately with the occurrence of a region of supersonic velocity, but, in the case of the thinnest section, is delayed from $M_\infty = 0.8$ nearly to $M_\infty = 0.9$. The worst effects occur with thicker sections. In the present experiments the airfoil with 12 per cent thick section actually suffered a reversal of the lift at a Mach number of 0.86. Some improvement in the aerodynamic characteristics can be achieved by refinement of the section shape. Much greater effects result, however, from changes in the planform of the wing. These will be taken up in the next article.

A,4. Effect of Sweepback.

INTRODUCTION. For the range of speeds within which the Kutta-Joukowski type of flow can be maintained, the straight wing of high aspect ratio shows the greatest aerodynamic efficiency. However, the advantage of this design depends in a critical way on the absence of pressure drag in the flow over the sections, and when the speed exceeds that for which this type of flow is possible the straight wing no longer appears to be the most suitable form. In the present article it will be shown that better efficiencies can be maintained by placing the wings at an oblique angle to the flow.

The use of sweepback to improve the efficiency of wings at supersonic speeds was first considered by Busemann [*32*] in 1935. Later, Betz, in Germany, pointed out that the critical Mach number of wings at subsonic speeds could be increased by the use of sweep. More recently it has been shown [*33*] that continuous flow of the Kutta-Joukowski type may be maintained at supersonic speeds and that the wave drag associated with such speeds disappears in the case of an infinite wing yawed at an angle greater than the Mach angle.

THEORY OF OBLIQUE CYLINDRICAL FLOW.

Principle of independence. The effect of sweepback is best visualized by considering the effect of sideslipping an infinitely long wing of constant cross section. Suppose the wing is initially in a stream with the oncoming velocity at right angles to the leading edge. It is clear that, if the flow is frictionless and all sections of the wing are alike, an axial or lengthwise motion may be introduced without causing any additional motion of the fluid, since the axial motion results only in a sliding of the surface parallel to itself. The combination of crosswise and lengthwise velocities results in an oblique relative motion. Conversely, in the oblique motion of a cylindrical wing the flow will be determined entirely by the crosswise, or normal, component of the velocity.

Since this principle of independence applies to any frictionless flow governed by the normal boundary condition, all the results of the two-dimensional flow theory can be applied immediately to the infinite oblique wing simply by subtracting the axial component of velocity. If the angle of sweep is large enough so that the crosswise component of the Mach number remains less than the critical Mach number of the wing sections, then the reversible subsonic type of flow over the sections can be preserved and the adverse effects of compressibility delayed to a higher speed. The maintenance of the essentially dragless subsonic type of flow at supersonic speeds requires that the wing be yawed at an angle greater than the Mach angle.

The elimination of drag at supersonic speeds is of course a feature of the two-dimensional flow. As later calculations will show, a drag does arise with wings of limited dimensions, chiefly as the result of departures from two-dimensional flow at the apex of the wing.

Further discussion of flow field. Since the effect of sweepback is of fundamental importance in the study of wings for high speeds, it will be worthwhile to consider it from several different points of view.

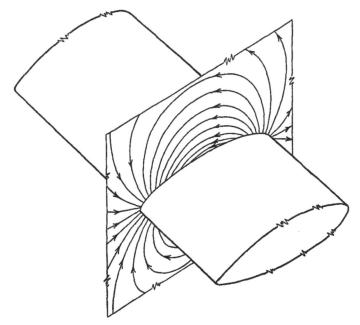

Fig. A,4a.   Velocities produced by wing in oblique motion.

If observed from a point at rest relative to the air, the motion produced by oblique translation of the wing would be confined in direction to planes cutting the wing at right angles to the leading edge (see Fig. A,4a). If observed from a point fixed relative to the wing, however, the velocity of flight would appear added vectorially to the disturbance velocities produced by the wing, and since the latter are generally small compared to the flight velocity, particles in crossing near the wing would be observed to depart but little from planes parallel to the flight direction.

With the wing at rest in an oncoming stream, the flow is steady and the cross sections of the individual stream tubes must follow the law of variation given by the one-dimensional flow equation (3-4). According to this equation the change in cross section of the stream tubes with changes in pressure bears a fixed relation to the local Mach number, independent of the shape of the body. However, the streamline pattern and the pressures around the sections of the yawed wing must be identical with those for a smaller Mach number, and the actual variation in stream tube area therefore must differ from the variation that appears in the flow over the sections. As shown in Fig. A,4b this difference appears in the plan view of the streamlines. Suppose for example that the oncoming stream has a supersonic velocity but the wing is yawed sufficiently so that the component of velocity normal to its leading edge is subsonic. Then the flow over the sections will have the configuration appropriate to the subsonic velocity, and the streamlines in this view will become narrower in the

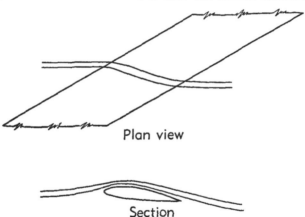

Plan view

Section

Fig. A,4b. Streamlines crossing oblique wing.

region of increased velocity and reduced pressure just above the upper surface of the wing. However, since the pressure gradients induced by the wing lie in a direction oblique to the main flow, the streamlines are curved laterally as they pass over the wing and this curvature has the effect of increasing the width of the stream tubes in the plan view in regions of excess velocity. Since the total velocity is supersonic, the resultant cross-sectional area must increase in these regions and hence the lateral widening observed in the plan view must predominate.[4]

The occurrence of shock waves and the consequent rapid rise in pressure drag on the oblique wing is to be expected when the crosswise component of the Mach number exceeds the critical Mach number of the

[4] It is of interest to note that the theory of oblique cylindrical flows provides a simple construction for the form of a channel required to produce the regular and continuous deceleration of a sonic or supersonic stream to a subsonic velocity. Starting with a subsonic two-dimensional channel flow, the addition of a constant velocity at right angles to the plane of the two-dimensional flow results in stream channels which show the required curvature and widening in the third dimension.

sections. It should be noted that the pressure force on an oblique wing of cylindrical form will act in the direction at right angles to its axis, and hence only a component of this force will be effective in the flight direction. Thus the pressure drag is reduced by yaw approximately in the ratio $\cos^3 \beta$, while the lift is reduced in the ratio $\cos^2 \beta$.

Relation to acoustic equation. The treatment of the infinite oblique wing by the thin airfoil theory is simplified by the solution of the acoustic

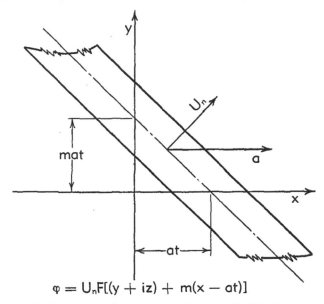

$$\varphi = U_n F[(y + iz) + m(x - at)]$$

Fig. A,4c. Wing flow represented by acoustic disturbance.

equation (3-9) in the form of the potential function

$$\varphi = F[(y + iz), (x + i\tau)] \tag{4-1}$$

(see [34]) where $\tau = iat$ and F is an arbitrary differentiable function of the two variables. The solution

$$\varphi = U_n F[(y + iz) + m(x - at)] = U_n F(\xi) \tag{4-2}$$

represents a disturbance of cylindrical form traveling with the velocity a along x, but with the normal velocity U_n (see Fig. A,4c). Differentiation for the velocities in the y, z plane results in

$$v = \frac{\partial \varphi}{\partial y} = U_n F'(\xi) \tag{4-3}$$

and

$$w = \frac{\partial \varphi}{\partial z} = U_n i F'(\xi) \tag{4-4}$$

If only the real solution is considered

$$\frac{v - iw}{U_n} = F'(\xi) = f(\xi) \tag{4-5}$$

and the function f may be selected with the aid of examples given in Table A,2. As required by the one-dimensional flow equation, the divergence of the velocity field is zero in planes at right angles to the direction in which the apparent velocity is the speed of sound (Fig. A,4c). By making use of the acoustic equation for the pressure, we obtain

$$\frac{\Delta p}{\frac{1}{2}\rho_\infty U_n^2} = \frac{2(\partial \varphi / \partial t)}{U_n^2} = \frac{f(\xi)}{\sqrt{1 - (U_n^2/a^2)}} \tag{4-6}$$

The formulas are of course independent of the resultant velocity U, which is not shown on Fig. A,4c, since it may lie at any angle β to U_n. If we introduce $U_n = U \cos \beta$ and $U_n/a_\infty = M_\infty \cos \beta$, there results

$$\frac{\Delta p}{\frac{1}{2}\rho_\infty U^2} = \frac{\cos^2 \beta}{\sqrt{1 - (M_\infty \cos \beta)^2}} f(\xi) \tag{4-7}$$

Applying formula 3 of Table A,2 and integrating for the total lift, one obtains

$$C_L = \frac{2\pi\alpha \cos \beta}{\sqrt{1 - (M_\infty \cos \beta)^2}} \tag{4-8}$$

for the lift coefficient of a thin oblique airfoil in two-dimensional flow.

It will be noted that the derivations just given furnish a simple graphical relation between the compressible and incompressible flow fields around a thin airfoil whether in straight or oblique motion. The resultant perturbation velocity vectors lie in the plane of the normal velocity U_n. Projection of these velocity vectors on the y, z plane yields the corresponding incompressible flow pattern as given by the Prandtl-Glauert transformation (see Fig. A,3b).

Experimental verification. Experimental determinations of the pressure and force characteristics of oblique wings show substantial agreement with the foregoing theoretical concept when allowance is made for the fact that perfectly two-dimensional flows cannot be realized experimentally.

Fig. A,4d shows the results of drag measurements made by the NACA on straight and sweptback wings attached to a falling body. The wings were attached internally to electrical balances and measurements of the drag force acting on the wings were transmitted to the ground by radio during the free fall [35]. The bodies were released from an altitude of approximately 40,000 feet and accelerated to a Mach number of nearly 1.3 before impact. The drags of the forward and rear wings differed slightly and the curves plotted on the figure are average values from the two sets of data. The results show the large gain predicted by the theoretical considerations. The departure of the experimental curve for the sweptback wing from the two-dimensional theory is seen to be satisfactorily explained by three-dimensional flow considerations. More detailed tests to check the theory were made in the wind tunnel of the Institute for Gas Dynamics, Braunschweig, Germany [*36*]. In these experiments a model (Fig. A,4e) which completely spanned the tunnel and which could be set at angles of yaw up to 40° was used. Pressure measurements

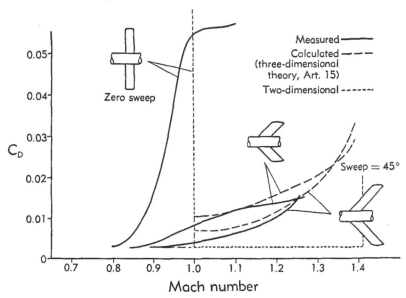

Fig. A,4d.   Drag coefficients of straight and swept wings, determined in free-fall experiments.

were made through a set of orifices spaced around the midsection of the airfoil. The results are shown in Fig. A,4f, A,4g, and A,4h.

Fig. A,4f and A,4g show the similarity of the pressure distributions obtained when the speed of the tunnel and the angle of yaw were adjusted to maintain a constant value of the crosswise velocity component. The last two pressure distributions, taken at a crosswise velocity component corresponding to $M_\infty \cos \beta = 0.77$, indicate the existence of a shock wave on the upper surface at about 40 per cent of the chord from the leading edge.

In Fig. A,4h the variations of lift curve slope with Mach number for the three angles of yaw are compared with the variations given by linear theory. In these experiments the angle of attack α was measured as the angle of rotation of the airfoil about its axis, and the variation of the lift

curve slope is therefore given by

$$\left(\frac{dC_L}{d\alpha}\right)_{M_\infty,\beta} = \left(\frac{dC_L}{d\alpha}\right)_{0,0} \frac{\cos^2 \beta}{\sqrt{1 - (M_\infty \cos \beta)^2}} \tag{4-9}$$

rather than by Eq. 4-8.

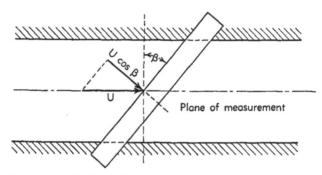

Fig. A,4e. Wind tunnel experiments with oblique airfoil [36].

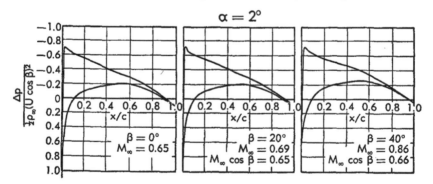

Fig. A,4f. Pressure distributions on oblique airfoil at $M_\infty \cos \beta = 0.65$.

OBLIQUE FLOWS WITH VISCOSITY.

Infinite cylindrical wing. The foregoing discussion has been based on the assumption of frictionless flow. Consideration of the effects of viscosity in oblique flows shows, however, that the most important relations for the yawed wing remain unmodified. In particular it can be shown that so long as the linear law of viscous stress usually adopted in the Navier-Stokes equations [1, pp. 571–577] remains valid, the pattern of the cross flow over an infinite oblique cylinder remains independent of

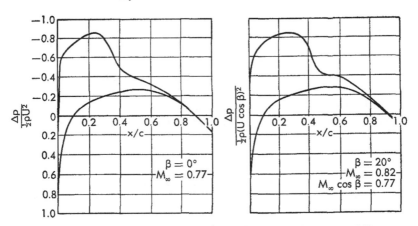

Fig. A,4g. Pressure distributions on oblique airfoil at $M_\infty \cos \beta = 0.77$.

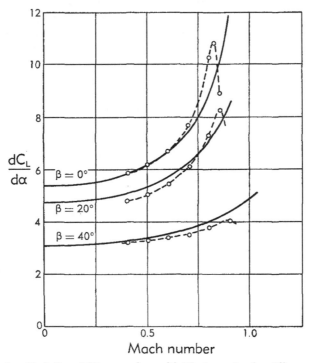

Fig. A,4h. Variation of lift curve slope with Mach number for oblique airfoils.

the axial motion of the cylinder [*37,38,39*]. This law of independence may be verified by writing the Navier-Stokes equations for a three-dimensional distribution of velocities and introducing the assumption that the form of the velocity field is independent of the axial position along the cylinder. The two equations for the cross flow then become independent of the third equation and are identical with the two-dimensional equations applicable to an unyawed cylinder traveling at a reduced velocity and reduced Reynolds number. The third equation remains dependent on the other two, however, and hence it is concluded that the cross flow is independent of the axial flow, but the axial flow is not independent of the cross flow.

The question of how far the principle of independence can be utilized depends on the range of validity of the linear stress relation. At very high supersonic speeds a shearing motion in one direction increases the temperature of the fluid and hence in effect modifies the coefficient of viscosity, so that the shearing stress in other directions is modified and a strict independence cannot hold. In the case of the turbulent boundary layer it seems that the nature of the internal stresses is not sufficiently well understood to establish the independence from first principles. The theory appears to be borne out, however, in the case of the flow over an oblique circular cylinder, which involves a large turbulent wake. Fig. A,4i shows the variation with yaw of the cross force developed by a circular wire. Since the test conditions entailed a range of crosswise Reynolds numbers in which the drag coefficient is known to be constant, the theoretical variation of the cross force coefficient in this case is almost exactly as $\cos^2 \beta$. The theoretical variation is shown in the figure by the solid line, while the experimental values from [*40*] are indicated by circles.

Finite aspect ratio. The theory of the action of viscosity in oblique flows leads to conclusions of considerable practical importance in the design of wings. First, it is to be noted that the thickness and velocity profile of the laminar boundary layer on an oblique wing are determined by the crosswise component of velocity alone. In the case of thin wings, experiment shows that the stalling angle, or angle of maximum lift, is determined by the phenomenon of laminar separation. If a straight wing and an oblique wing are now compared, it is seen that laminar separation can be expected at the same values of the lift coefficient and Reynolds number if these are based on the crosswise component of velocity. Thus, if a long narrow wing is placed at increasing angles of yaw β, the value of the lift at which flow separation takes place can be expected to diminish with $\cos^2 \beta$. With a sufficiently long wing the maximum lift can be expected to diminish in the same ratio. Such a wing thus behaves in nearly every respect as if it were flying at the reduced velocity $U \cos \beta$.

With sweptback wings of normal proportions it is observed experimentally that the relation between flow separation and maximum lift is not the same as it is with the straight wing. This difference can be understood if it is remembered that the similarity of flows has been demonstrated only in connection with the cross flows, or in other words with the flows observed by looking in the direction of the long axis of the wing. Separation occurs when a substantial volume of fluid adjacent to the upper surface of the wing is brought to rest or reversed in direction by the retarding action of the pressure field. Since the isobars are nearly parallel to the axis of the wing they can have little effect on the component of velocity parallel to the axis. Hence, at the point of flow separation on the

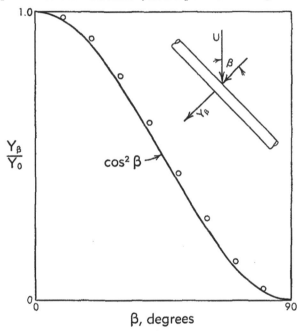

Fig. A,4i.   Theoretical and experimental variations
of the cross force on a circular wire.

oblique wing, the fluid merely loses the momentum carrying it toward the trailing edge—and flows in the direction parallel to the axis with the approximate velocity $U \sin \beta$ (see Fig. A,4j). It will be evident that once this endwise flow is developed on a wing of limited length the flow can no longer be approximated by that over an infinite wing. The separated air over the sections of the finite wing escapes at the downstream tip and hence does not remain to fill out an extended region or wake behind each section, such as appears behind the sections of a straight wing. The similarity of the flows can be expected to hold in practice for wings of normal proportions only so long as the boundary layer remains thin. After separation the similarity is lost.

As observed experimentally, the result is that flow separation on the straight wing is followed by a sharp drop in lift, whereas the lift of a wing with yaw or sweep of as much as 45° continues to rise with increasing angles of attack. With more extreme sweepback and especially with wings of low aspect ratio, the lift has been observed to increase up to angles of attack of 45°.

Fig. A,4k, taken from experiments made in the 40 by 80 ft wind tunnel of the Ames Aeronautical Laboratory, NACA, shows the results of

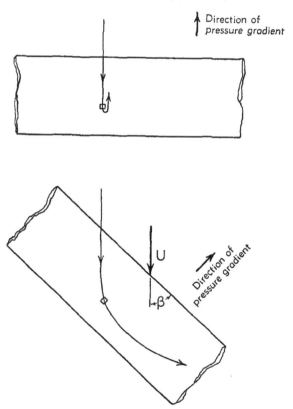

Fig. A,4j. Effect of yaw on the path of a particle in a separated boundary layer.

force measurements made on a model having approximately 60° of sweep. The tests were made at low speed, but the Reynolds number was of the order of 8 million. It will be noted that the drag curve follows the law of variation indicated by induced drag theory, viz.

$$C_D = C_{D_{min}} + \frac{C_L^2}{\pi R} \tag{4-10}$$

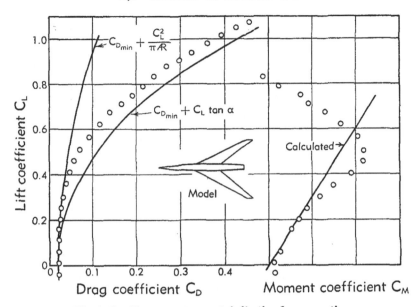

Fig. A,4k. Force measurements indicating flow separation on model with large angle of sweep.

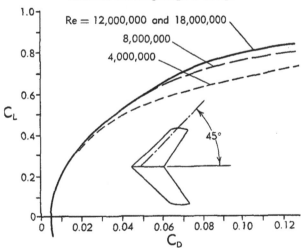

Fig. A,4l. Effect of Reynolds number on drag of swept wing, from [41].

up to a lift coefficient of approximately 0.3. Beyond this point the drag curve bends over and begins to follow the curve defined by

$$C_D = C_{D_{min}} + C_L \tan \alpha \qquad (4\text{-}11)$$

which is the variation expected when the resultant force is at right angles to the chord plane of the model and when no leading edge suction force is developed. Evidently flow separation occurs at about $C_L = 0.3$. This value corresponds to a lift coefficient of 1.2 with respect to the normal component of the stream velocity and hence is about equal to the normal stalling lift coefficient in straight flow. The effect of Reynolds number in delaying the onset of separation is shown in Fig. A,4l.

Further evidence of a change in type of flow is given by the moment curve shown on the right of Fig. A,4k. Although the lift continues to rise after flow separation occurs, there is some doubt that the higher lift coefficients can actually be used in practice because of the large increase in drag and the undesirable shift in center of pressure.

CHAPTER 2. AERODYNAMICS OF THIN WINGS AT SUBSONIC SPEEDS

A,5. Extension of Thin Airfoil Theory to Three Dimensions.

Necessity for more exact treatment of three-dimensional flow. The transition in high speed design to wings of sweptback planform and, in certain cases to wings of low aspect ratio, makes the necessity for three-dimensional treatment of high speed wing problems obvious. However, even in the case of straight wings, the approximation of two-dimensional flow becomes inadequate when higher speeds are considered. As explained in Art. 3, an increase of Mach number in the subsonic range leads to an increase in the magnitude and the vertical extent of the flow disturbance produced by the wing sections. Near the critical Mach number the vertical dimension of the region of disturbance becomes large relative to the chord of the wing, and, unless the aspect ratio is very great, the scale of magnitude of the section flows will become appreciable in terms of the wing span. As will be shown later, in the vicinity of sonic speed the flow over a lifting surface of finite span begins to take on the aspect of a two-dimensional flow in planes *normal* to the direction of flight.

Role of small disturbance theory in practical wing calculations. At the present time, calculations of wing flows in three dimensions are carried out almost exclusively on the basis of the linearized theory. Certain results of this theory, such as the well-known methods for determining the spanwise distribution of lift and the induced drag of wings at low speeds, have gained wide acceptance in aeronautical engineering.

The small disturbance theory represents mathematically the limiting case of infinitesmal thickness and angle of attack. The question of its applicability in practice depends on the magnitudes of these quantities as determined by practical considerations. At low speeds, increases of thickness or angle of attack are not necessarily accompanied by a pressure drag so long as flow separation is avoided. Hence in the design of airplanes for low speeds the thickness ratios are determined, broadly speaking, by the necessity for avoiding flow separation. Usually it is found that the resulting shapes are too thick to allow determination of the pressure distribution in detail by the small disturbance theory, although the lift curve slope, the center of pressure, and the induced drag are ordinarily given with sufficient accuracy by the linear relations.

At high subsonic speeds and at supersonic speeds, on the other hand, there arises a pressure drag or a wave drag which increases rapidly with the thickness ratio. For efficient flight at these speeds, therefore, it appears necessary to restrict the thickness ratio, or the slopes of the surfaces, to low values. In practice this will usually mean extending the dimensions of the wing or body in the flight direction. Of course this process leads to more unfavorable ratios of exposed surface area to span, or of surface area to volume, and hence increases the magnitude of the skin friction. Rough calculations can be made, however, which show that, in order to obtain reasonably small values of the total drag, extremely slender shapes must be used.

As an illustration of the change in proportions brought about by transition to higher speeds it is instructive to consider the case of a body of revolution designed to contain a given volume with minimum total drag. As is well known, at low speeds the minimum drag with friction occurs at a fineness ratio of about 3 to 1. At supersonic speeds, however, conservative estimates show that for the total drag to be a minimum the fineness ratio must be in excess of 20 to 1. As indicated in Art. 3 and in Fig. A,3c and A,3d, similar proportions must apply to the wing sections as well. Hence it seems reasonable to suppose that the small disturbance theory will provide useful engineering approximations for such bodies and wings as are actually well adapted to flight at high speeds.

Extension of the Prandtl rule to three-dimensional flows. The linearized treatment of wing flows in three dimensions makes use of (1) the approximate boundary condition of the thin airfoil theory and (2) the Prandtl transformation, which reduces the subsonic compressible flow to an equivalent incompressible flow. According to the Prandtl transformation, a distribution of φ's satisfying the differential equation of the compressible flow (3-8) can be made to satisfy Laplace's equation by merely lengthening or stretching the distribution along x by the factor $1/\sqrt{1 - M_\infty^2}$. It was seen that in two dimensions the flow over a thin airfoil section of unit chord length could be made to correspond to the incompressible flow over a geometrically similar section with its chord length increased by $1/\sqrt{1 - M_\infty^2}$. In this relation the spacing of the φ's in the vertical direction is not changed, so that $\partial\varphi/\partial z$ and hence the slopes of the streamlines at corresponding points remain unchanged. The crowding of the φ's in the x direction, however, leads to an increase in the perturbation velocity $u = \partial\varphi/\partial x$.

These same considerations are directly applicable to the three-dimensional wing flow.[5] Although the airfoil sections of the corresponding wings remain geometrically similar, the aspect ratios of the wings differ because of the change in chord length. The compressible flow over a wing of aspect ratio R at the Mach number M_∞ is related to the incompressible flow over a wing of aspect ratio $R\sqrt{1 - M_\infty^2}$. This relation between the wing shapes is illustrated by Fig. A,5. A study of the changes in the field of velocities and pressures around a given wing with increasing Mach number is thus reduced to the investigation of the incompressible flow around each of a series of wings of progressively reduced aspect ratio. Corresponding points in the compressible and incompressible flows are points of equal values of φ. These points occur at equal percentages of the chord lengths in the x-wise direction and at the same absolute values of y and z. The values of u, v, and w calculated for points in the field surrounding the extended wing are transferred to the corresponding points in the field

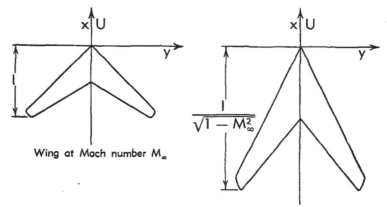

Fig. A,5. Wings having flows related by Prandtl transformation.

surrounding the actual wing with the values of u increased by the factor $1/\sqrt{1 - M_\infty^2}$. The field of the lengthened wing extends laterally and vertically to greater distances. Thus this process clearly shows the expansion of the field arising from compressibility.

As the Mach number approaches 1, the transformed wing becomes infinitely long in comparison with its width, and the flow becomes essentially two-dimensional in planes at right angles to the direction of flight. This consideration has lead to the adoption of the so-called "slender wing theory" (Art. 8) as a theory for sonic speeds.

Linearization of the boundary conditions in three-dimensional flows. As in the case of two-dimensional motion, the expression of the boundary

[5] For the treatment of more general bodies in three-dimensional flow, however, the reader should consult Göthert's paper [42].

condition in the form

$$\frac{dz}{dx} = \left(\frac{w}{U}\right)_{z \to 0}$$

results in a linear relation between the perturbation velocity field and the ordinates of the wing shape, so that again any wing shape may be divided into components in an arbitrary manner and its velocity flow fields may be determined as the sum of those arising from the individual components of the shape. Since problems associated with the camber and angle of attack of the mean surface present a somewhat different mathematical aspect from those associated with the thickness distribution, it is convenient to divide the wing into a thin inclined cambered or twisted "lifting surface" and a distribution of thickness which will be symmetrical above and below the plane $z = 0$.

As in the two-dimensional case, the distribution of thickness is characterized by equal and opposite values of the vertical velocity w on the upper and lower sides of the chord plane within the boundary of the wing planform, and can be represented by the action of a distribution of sources and sinks over the chord plane. Because of symmetry, the value of w is zero at all points of the plane $z = 0$ beyond the edges of the planform. The distribution of the velocities u and w extends continuously throughout the field. The thickness distribution alone gives rise to no lift and, in steady motion at subsonic speeds, to no drag other than that arising from the skin friction.

The flow associated with the lifting surface in three dimensions shows a discontinuity in v as well as in u across the chord plane. In steady motion the lifting pressure, and hence the discontinuity in u, vanishes outside the wing planform. The whole field of motion around the lifting surface can be obtained from the action of a suitable distribution of vortex filaments over the chord plane, and these filaments will extend behind the wing, forming the trailing vortex sheet. Since the trailing vortices lie parallel to the direction of flight, they represent a discontinuity in the lateral velocity v. The downwash velocity w remains continuous throughout the field.

It is important to note that the approximate boundary condition of the thin airfoil theory is limited, in three dimensions, to bodies that are thin and flat in all vertical cross sections. In such cases the flow, if continued to the interior of the body, will show a discontinuity of the velocities over an approximately plane mean surface which extends nearly to the edges of the planform. In the vicinity of such a planar distribution, the velocities and the slopes of the streamlines do not change rapidly with distance except possibly near the edges, and the condition of tangential flow can be applied without specification of the exact position of the solid boundary. Such a linearized boundary condition cannot be applied to other forms of slender bodies, however, even though the linearized flow equation may remain valid. Thus in the case of a body of revolution it is found that the streamlines of the internal flow emanate from a line of singularities along the axis of the body with an infinite radial velocity. The slopes of the streamlines diminish rapidly with the radial distance from the line, becoming equal to the slope of the surface only at the radial position of the surface. The range of physical validity of the flow equation is of course generally limited to the external flow field.

A,6. Lifting Surfaces.

GENERAL PROPERTIES.   Before proceeding to a discussion of specific wing forms it is desirable to consider certain results of the theory of lifting surfaces which are of a general nature. It is found that the linearized frictionless flow theory provides at once certain theorems relating to the lift and drag of thin wings. These theorems are independent of the particular planform of the wing and in certain cases they are also independent of the state of motion, i.e. the motion may be steady or unsteady and the speed may be subsonic or supersonic.

Drag of a given distribution of lift.   In general it is found that the lift of a wing is given quite accurately by considerations of frictionless flow. The drag, on the other hand, depends substantially on both the friction stresses and the distribution of normal pressures over the wing. However, for the determination of that part of the drag arising from the normal pressures, the frictionless flow theory may be used and an estimate of the total drag may be formed by adding an appropriate skin friction coefficient.

For the derivation of the theorems it is necessary to examine the process of computing the drag in the linearized theory. First consider a lifting surface having a planform such as shown in Fig. A,6a and a specified distribution of lift. The given distribution of lift will induce a certain distribution of downwash $w(x, y)$ over the plane of the wing, which results in a slope of the stream surface $dz/dx = w/U$. The form taken by the stream surface inside the boundary of the wing is identified with the form of the camber and twist of the surface required to produce the given distribution of lift. Since the drag with which we are concerned arises only from the normal pressures, it can be computed by multiplying the lift of each point of the planform by the rearward inclination of the surface at that point and integrating the resultant product over the entire wing[6] (see Fig. A,6a).

Computation of the drag of a given distribution of lift thus depends first on the computation of the downwash at each point. In a linearized

---

[6] If the given distribution of lift shows infinite values around the leading edge, the calculation must be performed over a stream surface slightly above the chord plane and the integration must extend a small distance beyond the wing boundary, as explained in Art. 2.

theory this may be accomplished by dividing the wing up into elementary areas and considering the field of downwash induced by each lifting element. The pattern of the downwash field surrounding an element of lift will be determined by an elementary solution of the flow equation, which may be elliptic or hyperbolic. In any event the downwash pattern will be the same for every element except for a constant factor of strength proportional to the lift of the element.

Determination of the downwash at any given point of the planform requires an integration of the effect at this point of all lifting elements of

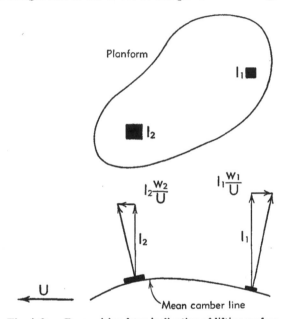

Fig. A,6a.   Drag arising from inclination of lifting surface.

the wing. One integration over the surface is required to yield the drag of one element of lift, and a second surface integration will be required to yield the total drag.

EQUALITY OF DRAG IN REVERSED MOTION.   A consideration of the elements of this double integral leads to the conclusion that *the drag of a given distribution of lift is unchanged by a reversal of the direction of motion.* Each element of the drag involves the product of the lift of one element l_1 and the downwash induced at its position by another element l_2. This downwash may be written

$$w_{12} = W_{12}l_2$$

where W_{12} denotes the downwash at position 1 due to a unit lift at position 2. The element of drag is then

$$l_1 W_{12} l_2$$

Now, if the direction of motion is reversed, element 2 will lie in the same relation to element 1 as was formerly occupied by l_1 relative to l_2. In other words the influence function W_{21} in reversed motion will be equal to W_{12} in forward motion. The drag of l_1 in the downwash of l_2 in forward motion is then equal to the drag of l_2 in the downwash of l_1 with the motion reversed. In symbols,

$$l_1 W_{12} l_2 = l_2 W_{21} l_1$$

For every element of drag there is thus a corresponding equal element for motion in the reverse direction. Since the correspondence includes all elements of the double integrals, the total drags are equal (see Fig. A,6b).

The invariance of drag evidently applies under very general conditions. The elements of lift need not lie in the same, nor in parallel, planes. At supersonic speeds or in unsteady motion at subsonic speeds, the zones of interaction between elements may be limited in extent. Reversal of an unsteady motion implies reversal of the velocity vector $\mathbf{q}(t)$ at each corresponding instant. For the case of steady motion at subsonic speeds the equality of forward and reverse drags may be shown by Munk's stagger theorem [7]. Von Kármán [43] and Hayes [44] have demonstrated the theorem for the case of steady supersonic motion.

Distribution of Lift for Minimum Drag.

Mutual drag in combined flow field. Certain necessary conditions on the distribution of lift for minimum drag can be derived through the consideration of the fields of disturbance velocities produced by a given distribution of lift in forward and reversed motion. The derivation makes use of the idea, originally advanced by Munk [45], of superimposing these velocity fields.

Let us first suppose that the planform of a wing is given, together with the distribution of lift. If the same distribution of lift is considered in reversed flow, it will in general be found that the downwash distribution is different from the original one, and that the wing must therefore be given a different distribution of camber and twist to maintain the same distribution of lift. In spite of this difference in shape, however, the total drag must remain unchanged by the reversal. Since the drag is obtained by integrating the product of the lift and the downwash over the planform, we have

$$D = \frac{1}{U} \iint l(x, y) w_f(x, y) dx dy = \frac{1}{U} \iint l(x, y) w_r(x, y) dx dy \quad (6\text{-}1)$$

where w_f and w_r are the downwash distributions in forward and reversed flow respectively.

The drag may be computed by considering the given distribution of lift to be placed in a "combined flow field," obtained by superimposing the perturbation velocities for forward and reversed motion. It is evident that for the same distribution of lift to appear in each field the horizontal perturbation velocities u must be exactly equal and opposite at each point

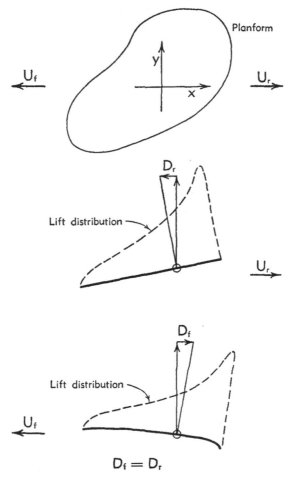

Fig. A,6b.   Drag of a given distribution of lift in forward and reversed motion.

of the wing surface. Hence the discontinuity in u vanishes in the combined field. A continuous distribution of downwash remains, however. If its value is denoted by

$$\bar{w} = w_t + w_r$$

the drag will be given by

$$D = \frac{1}{2U} \iint l(x, y)\bar{w}(x, y)dxdy \tag{6-2}$$

The quantity $\overline{w}/2$ may be interpreted as the mean value of the downwash for the two directions of motion and reduces to the induced downwash of the Prandtl wing theory in the case of steady motion at subsonic speeds.

It may now be shown that *the mutual interference drags of two distributions of lift in the combined flow field are equal.* Consider two elements of lift such as l_1 and l_2 in Fig. A,6c. The drag of l_1 caused by the downwash of l_2 in the combined flow may be written

$$l_1 \overline{W}_{12} l_2$$

where the symbol $\overline{W}$ denotes the combined downwash arising from a unit element of lift. Similarly, the drag of l_2 caused by the interference of l_1 is given by

$$l_2 \overline{W}_{21} l_1$$

From the symmetry of the combined flow field for each element it is evident that

$$\overline{W}_{12} = \overline{W}_{21}$$

and hence the interference drags are equal. This equality of mutual drags holds for groups of elements and hence for any two distributions of lift.

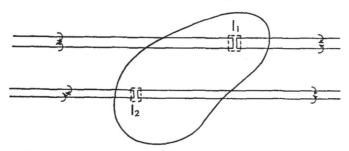

Fig. A,6c.   Mutual interference of elements in combined flow field.

Distribution of lift for minimum drag with a given planform and a given total lift.   Suppose the planform of the wing to be given, together with a distribution of lift designed to achieve the minimum drag for a given total lift L. If the drag is actually a minimum, then a small change in the shape of the distribution will cause no first order change in the drag. The change in shape may take the form of a small distribution added to the original lift; the condition of a fixed total lift will be met by admitting only those variations for which the total lift is zero. Such restricted variations may be divided into pairs of equal, opposite elements of lift, of which one pair may be taken as representative. Following the law of superposition, the drag added by such a pair will be composed of three parts:

1. the drag of the added elements acting alone
2. the interference drag of the added elements caused by the downwash of the original distribution

3. the interference drag of the original distribution caused by the downwash of the added elements.

The first item will be of second order in terms of the small magnitude of the added distribution. (The fact that this second order term is always positive insures that a stationary value of the drag will be a minimum and not a maximum.) According to the mutual drag theorem, the second and third items will be equal in the combined flow field. The first order variation in drag is then equal to twice the drag of the added lift in the downwash of the original distribution. For a pair of equal and opposite lifting elements the drag will be zero if the downwash is the same at each element. Furthermore, the drag added by every such pair will be zero if the downwash has the same constant value at each point of the planform.

Hence if a distribution of lift is to have the minimum drag consistent with a given total lift the "combined" downwash $\bar{w}$ must be constant over the entire planform (see Fig. A,6d).

Planform for minimum drag. The foregoing condition for minimum drag may be generalized to include variations in planform of the wing, provided the meaning of the term "additional lift distribution" is broadened to cover small changes in the outline shape of the planform. It will be apparent from the foregoing discussion that the drag for a specified total lift will be a minimum with respect to extensions of the wing outline in a given direction if the downwash $\bar{w}$ is not only constant over the planform but maintains this value without first order change for a small distance beyond the border of the planform. If the distribution of lift over the planform is plotted as a surface above the chord plane, there will result a three-dimensional figure whose volume is the total lift. *Then for the drag to have a stationary value relative to small changes in shape of this figure, $\bar{w}$ must be a constant over the planform and if s denotes a distance in the x, y plane measured from the edge of the planform $d\bar{w}/ds$ must equal zero.*

Such general variations in shape of the lift distribution do not ordinarily lead to the determination of a definite minimum of drag since at subsonic speeds, with the total lift fixed, the drag decreases monotonically with increases of span, while at supersonic speeds it decreases indefinitely with increases of both span and chord. The foregoing conditions are useful, however, in determining the minimum drag consistent with certain limitations on the planform, as later examples will show.

Relation to theory of induced drag. For the case of steady motion at subsonic speeds the foregoing theory agrees with the classical theory [7,8] of induced drag. In steady motion no wave drag arises and the whole pressure drag can be associated with the downwash produced by the trailing vortex wake. In the superposition of the forward and reversed flow fields for a fixed distribution of lift, wakes having identical distributions of vorticity will extend both ahead of and behind the wing. Since the

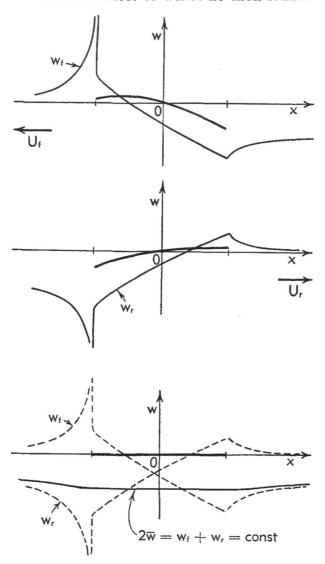

Fig. A,6d.   Condition for minimum drag with a given total lift.

longitudinal component u of the perturbation velocity is cancelled at each point within the planform boundary, the vortices in the superimposed flow will continue straight across the wing in a direction parallel to the stream velocity (see Fig. A,6c). In the subsonic case the superimposed flows form a two-dimensional field of motion associated with a vortex ribbon or wake of width equal to the span and extending from $+\infty$ to

— ∞ in the stream direction. The distribution of vorticity in the combined wake is identical with the distribution in each of the single semi-infinite wakes and bears the same relation to the spanwise distribution of lift as given by the classical wing theory. The minimum drag for a given planform and a given total lift is achieved when the spanwise distribution of lift is elliptical—since in this case the downwash $\overline{w}$ induced by the combined wakes is constant—and is given by the well-known formula

$$D = \frac{L^2}{\pi \frac{1}{2}\rho_\infty U^2 b^2}$$

where b is the span of the lifting surface. Inasmuch as the constant value of $\overline{w}$ extends not only over the wing planform, but also over the whole infinite vortex ribbon ahead of and behind the wing, the drag remains unchanged for extensions of the planform in the chordwise direction.

According to Munk's stagger theorem, the induced drag is unaffected by the form of the distribution of lift in the chordwise direction, even though the spanwise distribution does not correspond to the minimum drag. The present considerations can be seen to yield this result by noting that in the combined flow field, which is a two-dimensional motion, $\overline{w}$ remains independent of x even though a spanwise variation is admitted. Variations which amount to a redistribution of lift along any line in the flight direction thus cannot change the drag since the downwash $\overline{w}$ has the same value at all points along such a line. (This stationary character of the drag is, of course, not that of a minimum value since even finite changes of chordwise lift distribution produce no change of drag.)

EFFECT OF COMPRESSIBILITY ON DRAG AT SUBSONIC SPEEDS. An important result of the linearized wing theory is the prediction that *the drag associated with a given distribution of lift is unchanged by changes of the Mach number in the subsonic range.* This result can be verified by applying the Prandtl transformation to the combined flow field. In this field the potential φ has the same value at all points of any line parallel to the x axis. A change in the scale of x thus produces no change in the gradient of the potential and hence does not alter the perturbation velocities. The drag which, it will be remembered, results from an integration of the product of the given lift and the downwash of the combined field at each point is thus also unchanged.

If the geometry and angle of attack of the wing are given rather than the lift distribution, then the spanwise distribution of lift and the induced drag will be affected by the Mach number, though the change may be slight. According to the Prandtl rule (Art. 5), the flow pattern at increased Mach numbers can be related geometrically to the incompressible flow around a derived wing of increased chordwise dimensions or reduced aspect ratio. The computation of induced drag at varying Mach numbers thus involves first of all the computation of the varying lift distribution.

In many cases the lift distribution is nearly elliptical and, as will be shown later, tends to become more so at higher subsonic Mach numbers so that the variation of drag with a given total lift is slight.

REVERSAL THEOREM FOR LIFT CURVE SLOPE. The combination flow, obtained by superimposing the disturbances arising in forward and reversed motion, has been used by Brown [46] to show that the lift curve slope of a wing in steady motion is unchanged by a reversal of the planform. Here it is sufficient to show that the lift of a flat surface at an angle of attack α is unchanged (except for sign) by a reversal of the flow direction, since, in the linear theory, the lift curve slope is unaffected by a small initial camber or thickness. The distribution of lift over the planform will, in general, be different for the two directions of motion. The flow in forward motion must satisfy the boundary condition $\overrightarrow{w} = U\alpha$ at every point of the planform. With the wing fixed and the flow reversed, the boundary condition requires that the downwash at each point be equal and opposite to that in forward flow; i.e. $\overleftarrow{w} = -\overrightarrow{w}$. Superimposing the two flows then results in a complete cancellation of downwash at each point of the planform, and the combined flow field thus satisfies the boundary condition for a flat plate at zero angle of attack.

The proof of the reversibility of the lift requires consideration of the mechanism of the drag, although the drag itself does not in general remain unchanged by the reversal of the planform. The drag in forward motion will be given by

$$\overrightarrow{D} = \overrightarrow{L}\alpha - \overrightarrow{T}$$

where the quantity $\overrightarrow{T}$ is introduced to take into account the leading edge thrust and is proportional to the strength of the singularity in the velocity around the leading edge. The drag of the airfoil can also be obtained by a different consideration, which involves a characteristic feature of the three-dimensional wing flow. In such flows the disturbance vanishes at a great distance ahead of the wing; on the downstream side, however, the disturbance remains, in the form of the wake, undiminished by distance. At supersonic speeds a downstream disturbance appears also in the Mach waves emanating from the wing. In any event the drag appears as the rate of increase of momentum involved in the continual extension of this wake at the flight velocity and can be obtained from the configuration of the flow at a great distance behind the wing. From this reasoning it will be apparent that when forward and reversed flows are superimposed, the two wakes, which now lie on opposite sides of the wing, will not interfere, and the combination flow at a great distance from the wing in either direction will be identical with either one or the other of the single flows. If the uniform stream velocity U is now added to the combined flow and the change in momentum introduced at the wing is considered, it can be seen that the drag arising from the wake extending forward is simply the negative of the drag for reversed flow, while the drag of the wing in forward flow appears in the rear wake with its normal sign. The drag in the combination flow is thus the difference of the two drags; i.e.

$$\overline{D} = \vec{D} - \overleftarrow{D}$$

Now a second independent determination of the drag can be obtained from a consideration of the inclination of the pressure forces over the surface of the wing itself. This method of calculation leads to the expressions

$$\vec{D} = \vec{L}\alpha - \vec{T}$$

$$\overleftarrow{D} = \overleftarrow{L}\alpha - \overleftarrow{T}$$

where the terms $\vec{T}$ and $\overleftarrow{T}$ are again introduced to include leading edge thrust forces. Inasmuch as the combined flow corresponds to a flat horizontal wing surface, the only drag arising in this flow must be the resultant of the suction forces around the edge of the wing or, in other words,

$$\overline{D} = \overleftarrow{T} - \vec{T}$$

provided the suction forces T are not changed in value by the superposition. The latter condition is found to be satisfied if the Kutta condition is imposed on each flow, for then the infinite edge velocities of the two flows involve completely separated regions of the planform; the trailing edge for one direction of motion becomes the leading edge for the other direction and the tangential trailing edge velocities can be added to the higher order leading edge velocities without affecting the suction force. The theorem can now be proved by writing

$$\vec{D} - \overleftarrow{D} = (\vec{L}\alpha - \vec{T}) - (\overleftarrow{L}\alpha - \overleftarrow{T}) = \overleftarrow{T} - \vec{T}$$

and hence

$$\vec{L} - \overleftarrow{L} = 0$$

The reversibility of the lift curve slope applies to steady motion at subsonic or supersonic speeds without restriction on the planform. It obviously applies also to airfoils in combination, such as wing and tail configurations.

RECIPROCAL RELATIONS IN THIN AIRFOIL THEORY. In spite of the variety of form in three-dimensional wing flows, there thus exist certain remarkable and simple relations between the disturbance fields in forward and reversed motion. Consider, for example, the application of Brown's reverse flow theorem to the lift of a narrow triangular wing at supersonic speed. With the triangle moving point foremost, the flow in the vicinity of the wing is similar (Art. 8) to a solution of Laplace's equation in two dimensions, expanding smoothly in the downstream direction in proportion to the width of the triangle. With the triangle moving base foremost, the pressure distribution over the surface and the flow pattern in the vicinity of the wing have a decidedly different appearance. The flow is deflected at the leading edge by discontinuous wave fronts and the pressure distribution over the surface following is characterized by a crossed array of Mach lines along which ridges or discontinuous slopes in the pressure distribution appear. Nevertheless, according to Brown's theorem the total lift must be the same in the two cases.

A more general reverse flow theorem which includes the lift theorem as well as a number of other interesting relations is the one due to Ursell and Ward [47]. This theorem has recently been generalized by Heaslet and Spreiter [48] and by Flax [49], who include time-dependent as well as steady motions. In [48], the theorem is related to the classical reciprocal theorems in dynamics associated with the names of Maxwell, Helmholtz, and Rayleigh.

In the form given by Ursell and Ward, the theorem applies specifically to steady motion and may be stated in the following way: Consider a thin wing of planform S and let the slope of the surface, which is given arbitrarily, be α_1. These specifications together with the assumptions of the thin airfoil theory then determine a corresponding distribution of pressure p_1. Both α_1 and p_1 may vary from point to point over the surface and are not necessarily the same on the upper and lower sides of the wing. Now let the motion of the planform be reversed and consider a second independent distribution of slope α_2 together with its corresponding distribution of pressure p_2. The theorem is:

$$\int_S \alpha_1 p_2 dS = \int_S \alpha_2 p_1 dS \tag{6-3}$$

where the integration extends over both upper and lower sides of the wing. In order to make the relations connecting α_1 to p_1 and α_2 to p_2 definite, Ursell and Ward assumed that the Kutta condition was applied.

It is instructive to derive the relation (6-3) by the method previously used to demonstrate Brown's reversal theorem for the lift.[7] Supposing the range of integration to be limited strictly to the surface S, so that terms contributing to the leading edge thrust do not arise in the integration, we have for the drags in forward and in reversed motion,

$$D_1 = \int_S \alpha_1 p_1 dS + T_1 \tag{6-4}$$

$$D_2 = \int_S \alpha_2 p_2 dS + T_2 \tag{6-5}$$

---

[7] This general application of the method is contained in an unpublished work of W. D. Hayes.

where T_1 and T_2 denote the leading edge forces. Following Munk [45] we now superimpose the two disturbance fields and consider the combined field to be placed in a stream of velocity U. The angle of deflection in the resulting flow over the surface S will be $\alpha_1 + \alpha_2$, while the pressure will be $p_1 - p_2$. We then have for the drag in the combined disturbance field

$$D = \int_S (\alpha_1 + \alpha_2)(p_1 - p_2)dS + T_1 - T_2 \qquad (6\text{-}6)$$

Eq. 6-6 is based on the assumption that the edge forces T_1 and T_2 are additive, and this will be the case if the Kutta condition is applied, since the portions of the edge on which T_1 and T_2 may arise are then completely separated in the forward and reversed flow.

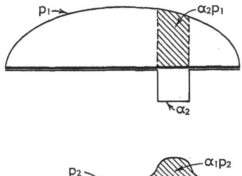

Fig. A,6e.   An application of the reciprocal relation (Eq. 6-3).

Now a second, independent determination of the drag in the combined disturbance field may be made by means of a momentum survey taken at a distance from the wing. Since the combined field contains disturbances extending in both directions from the wing, the momentum integral must include regions on the upstream side of the wing. However, because of the one-sided nature of each individual disturbance field, they do not interfere, and the superposition of the two fields produces no change in either of the components at a sufficient distance from the wing. The balance of momentum between regions far upstream and far downstream in the combined field then shows the drag to be simply the difference of the drags of the individual flows, or

$$D = D_1 - D_2$$

After expressing D_1 and D_2 by means of Eq. 6-4 and 6-5 and equating the result to Eq. 6-6 we obtain

$$\int_S (\alpha_2 p_1 - \alpha_1 p_2)dS = 0$$

which is the same as Eq. 6-3.

An important result of the foregoing relation, which commits itself easily to memory, may be stated as follows (see Fig. A,6e): *The lift on a wing due to deflection of a flap is equal to the lift on the flap due to deflection of the wing with the flow reversed.* The term flap may here be taken to mean any small area of the wing surface. The statement may be verified by taking for α_1 a unit value over the whole surface, and for α_2 a value equal to unity over the region of the flap but equal to zero over the remainder of the wing.

The foregoing relation shows immediately those regions of a wing on which a flap is most effective in producing lift. The regions of greatest effectiveness are just those regions on which the lift density is greatest for the flat wing with the flow reversed. Thus in the case of a flat triangular wing, moving point foremost at an angle of attack, it is known that the lift density approaches a high peak along the leading edge. If the triangle is reversed so as to travel base foremost, the leading edge is turned into a trailing edge and it follows that the effectiveness per unit area of a narrow flap along this edge approaches a similar high peak value.

A,7. Lifting Surfaces. Solutions for Specified Planforms.

THE ELLIPTIC WING AT SUBSONIC SPEEDS. Analytic solution of Laplace's equation in three dimensions is practicable only for simple geometric forms. Of the few completely determined solutions in three dimensions the most interesting from the standpoint of wing theory are the classical solutions for the potential of ellipsoids [1], which include the elliptic disk as a limiting form. A treatment of the elliptic wing on this basis has been given by Krienes [50], who generalized earlier work by Kinner [51] on the circular disk. The interpretation of the classical solutions in terms of lifting surface theory was suggested by Prandtl and depends on the consideration of the pressure as a potential field. Accordingly, Prandtl defines a function

$$\Psi = -\frac{\Delta p}{\frac{1}{2}\rho_\infty U^2} = -\frac{p - p_\infty}{\frac{1}{2}\rho_\infty U^2} \tag{7-1}$$

The function Ψ is a solution of Laplace's equation, and has been termed the "acceleration potential." In steady motion the pressure disturbance Δp is related to the horizontal perturbation velocity by $\Delta p / \frac{1}{2}\rho_\infty U^2 = -2u/U$; hence in this case Ψ becomes simply $2u/U$. The corresponding vertical component w/U, whose value in the plane of the wing determines the camber and twist of the lifting surface, is found by an integration of the vertical acceleration experienced by a particle, beginning with the parallel flow at infinity ahead of the wing and ending at the point where the downwash is desired. The integral to be evaluated is

$$w = \int_{-\infty}^{x} \frac{\partial u}{\partial z}\, dx \tag{7-2}$$

Following the classical potential theory, solutions are obtained in terms of the orthogonal ellipsoidal coordinates λ, μ, and ν. The surfaces λ-const are confocal ellipsoids, while the surfaces defined by μ and ν are

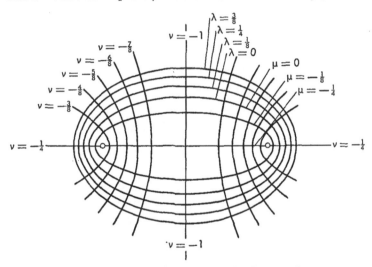

Fig. A,7a.   Ellipsoidal coordinates in plane $z = 0$.

intersecting hyperboloids. The ellipsoidal coordinates are related to the Cartesian coordinates by equations of the form

$$\frac{x^2}{a^2 + \lambda} + \frac{y^2}{b^2 + \lambda} + \frac{z^2}{c^2 + \lambda} = 1 \tag{7-3}$$

where a, b, and c are the semiaxes of the base ellipsoid. Thus the base ellipsoid is given by $\lambda = 0$ and a thin ellipsoid is obtained by letting the semiaxis c approach zero. Setting $\lambda = -c^2$ gives the focal disk, which lies in the chord plane and forms the surface of discontinuity of the velocities u or w when the velocity field is continued into the interior of the base ellipsoid. The intersections with the disk of the surfaces of constant μ form a family of confocal ellipses; lines of constant ν appear as a family of hyperbolas (see Fig. A,7a).

The class of solutions considered is that expressible as the product of three functions of the three ellipsoidal coordinates. For the elliptic wing, solutions are given by products of the form

$$2\frac{u_n^m}{U} = E_n^m(\mu)E_n^m(\nu)F_n^m(\lambda) \tag{7-4}$$

where E_n^m and F_n^m are Lamé's functions of the first and second kind, respectively, of order n and class m. The use of the function of the second kind F_n^m is dictated by the requirement that the pressure vanish at infinity. The value of F is of course constant on the wing surface, and the distribution of surface pressure is determined in shape by the product $E(\mu)E(\nu)$. The degree n of the function determines the number of alternations in sign of the pressure distribution over the wing, while the class of the function determines the orientation of the isobars relative to the x, y, and z directions. Examples of the first few Lamé functions are

$$E_0 = 1$$

$$E_1^1(\lambda) = \sqrt{a^2 + \lambda}$$

$$E_1^2(\lambda) = \sqrt{b^2 + \lambda}$$

$$E_1^3(\lambda) = \sqrt{c^2 + \lambda}$$

$$E_2^1(\lambda) = \sqrt{(b^2 + \lambda)(c^2 + \lambda)}$$

etc. Lamé products of lower degree bear simple relations to the Cartesian coordinates. Thus

$$E_1^1(\lambda)E_1^1(\mu)E_1^1(\nu) = \sqrt{(a^2 - b^2)(a^2 - c^2)}\, x \tag{7-5}$$

$$E_1^2(\lambda)E_1^2(\mu)E_1^2(\nu) = \sqrt{(b^2 - a^2)(b^2 - c^2)}\, y \tag{7-6}$$

$$E_1^3(\lambda)E_1^3(\mu)E_1^3(\nu) = \sqrt{(c^2 - a^2)(c^2 - b^2)}\, z \tag{7-7}$$

Functions of the second kind are obtained by the integration

$$F_n^m(\lambda) = E_n^m(\lambda) \int_\lambda^\infty \frac{d\lambda_1}{[E_n^m(\lambda_1)]^2 \sqrt{(a^2 + \lambda_1)(b^2 + \lambda_1)(c^2 + \lambda_1)}} \tag{7-8}$$

For a complete discussion of ellipsoidal coordinates and associated functions, the reader is referred to Whittaker and Watson [52], Hobson [53], or to textbooks on potential theory.

As in the case of the complex velocity functions (Table A,2), a distinction must be made between functions for which u is discontinuous across the chord plane ($\lambda = -c^2$) and those for which w is discontinuous. Functions of the latter type yield various distributions of thickness over the base disk. Of the functions yielding distributions of lift (u discontinuous), the simplest is (Fig. A,7b)

$$2\frac{u_1^3}{U} = E_1^3(\mu)E_1^3(\nu)F_1^3(\lambda) \tag{7-9}$$

which reduces to

$$\frac{u_1^3}{U} = zf(\lambda) \tag{7-10}$$

For an ellipsoid of vanishing thickness it is found that

$$\frac{u_1^3}{U} = \frac{z_{\text{ell}}}{c} = \sqrt{1 - \frac{x^2}{a^2} - \frac{y^2}{b^2}} \qquad (7\text{-}11)$$

The distribution of lifting pressure over the wing in this case can be represented by the ordinates of an ellipsoid constructed over the basic disk. By

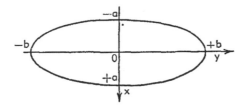

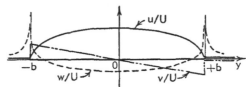

Velocity distribution along y axis

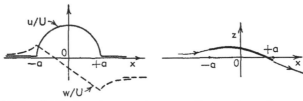

Velocity distribution along x axis   Section at $y = 0$

Fig. A,7b. Velocities given by first potential function for elliptic disk.

making use of the relation (7-11), the total lift of the distribution is found to be

$$L = \frac{8}{3}\pi a b \frac{1}{2}\rho_\infty U^2 \qquad (7\text{-}12)$$

or, in coefficient form,

$$C_L = \frac{8}{3} \qquad (7\text{-}13)$$

If the surface distributions of pressure, corresponding to the Lamé functions, are integrated in the chordwise direction to obtain the spanwise distributions of lift or circulation it is found that the latter are given by simple algebraic curves in the form of Legendre polynomials. Thus, if the total circulation around a section is expressed as $\Gamma(\eta)$, where $\eta \approx y/b$, it is found that

$$\Gamma_1^3(\eta) = 2\pi U a \frac{P_0(\eta) - P_2(\eta)}{3} = \pi U a(1 - \eta^2) \qquad (7\text{-}14)$$

and the spanwise loading is parabolic.

For the determination of the downwash it is found convenient to use the general form [52, p. 390] of the three-dimensional potential

$$2\frac{u}{U} = \oint F[(\alpha x + \beta y + \gamma z), t]dt \qquad (7\text{-}15)$$

Then the expression for w/U becomes

$$\frac{w}{U} = \oint \frac{\gamma}{\alpha} F[(\alpha x + \beta y + \gamma z), t]dt \qquad (7\text{-}16)$$

For the solutions represented by Lamé functions it has been shown [52, pp. 567–570] that

$$2\frac{u_n^m}{U} = \frac{1}{2\pi i} \oint Q_n(\alpha x + \beta y + \gamma z)E_n^m(t)dt \qquad (7\text{-}17)$$

with

$$\alpha = \frac{\sqrt{k^2 - t^2}}{ka}$$

$$\beta = \frac{t}{kb} \qquad \left. \right\} \quad (\alpha^2 + \beta^2 + \gamma^2 = 0)$$

$$\gamma = \frac{i\sqrt{1 - t^2}}{a}$$

and

$$k = \sqrt{1 - \frac{a^2}{b^2}}$$

Here Q_n is the Legendre function of the second kind. The contour c extends along the real axis with loops around the points $\pm k$. With $t/k = \sin\theta$, the integration (7-17) may be interpreted geometrically as the superposition of two-dimensional fields of the form given by formula 7 in Table A,2 at varying angles θ around the z axis. The downwash function is given by (see Eq. 7-16)

$$\frac{w_n^m}{U} = \frac{1}{4\pi} \oint k \sqrt{\frac{1 - t^2}{k^2 - t^2}} Q_n(\alpha x + \beta y + \gamma z)E_n^m(t)dt \qquad (7\text{-}18)$$

For the potential function of the first degree, evaluation of the integral yields (see [50])

$$\frac{w_1^3}{U} = \frac{a}{b} Q_1\left(\frac{y}{b}\right) - E(k)\frac{x}{a} \qquad (7\text{-}19)$$

within the boundary of the disk. Here $E(k)$ is the complete elliptic integral of the second kind, equal to the ratio of the semiperimeter of the elliptic disk to the span $2b$. The term $E(k)(x/a)$ yields a parabolic camber shape for the sections taken parallel to the flight direction, while the remaining term shows a progressive spanwise twist of the sections along the span. As the aspect ratio $R = 4b/\pi a$ approaches infinity, the spanwise variation $(a/b)Q_1(y/b)$ disappears, and the flow over the sections approaches in form the two-dimensional velocity function given by formula 5 of Table A,2.

Differentiation of Eq. 7-14 with respect to y yields the spanwise gradient of the circulation and hence the discontinuity of the lateral velocity v in the wake:

$$\frac{1}{U}\frac{d\Gamma_1^3}{dy} = 2\frac{v}{U} = -2\pi\frac{a}{b}P_1(\eta) = -2\pi\frac{a}{b}P_1\left(\frac{y}{b}\right) \qquad (7\text{-}20)$$

At a great distance behind the wing, the wake forms a two-dimensional field of motion, with the axis parallel to the flight direction. Substituting v for u in formula 9 of Table A,2 shows that the velocity field of the wake is given by

$$\frac{v - iw}{U} = 2i\frac{a}{b}Q_1\left(\frac{y}{b} + i\frac{z}{b}\right) \qquad (7\text{-}21)$$

The downwash along the centerline of the wing (Eq. 7-19, $x = 0$) is found to be equal to one-half the final value induced by the trailing vortices in the wake. As may be shown by considerations of symmetry, this statement must apply to every distribution of lift which is symmetrical about a line at right angles to the flight direction.

It is of interest to note that the foregoing relations for the ellipsoidal pressure distribution agree qualitatively with the results that would be obtained by application of the two-dimensional wing section theory together with the Prandtl lifting line theory. For the same camber and twist, however, the more complete calculation [54] shows a reduction of the lift in the ratio $1/E$.

Lift distributions of more general form may be obtained by expansion in a series of ellipsoidal harmonics, making use of the orthogonal properties of the Lamé functions. The distributions obtained in this way, however, fall to zero all around the edge of the wing, with tangential flow at both leading edge and trailing edge. In order to show the effect of changes in angle of attack of a rigid wing surface it is necessary to obtain solutions with flow around the leading edge; it is to be expected that such solutions will show infinite velocities around the edges of the focal disk.

Solutions having the desired properties may be obtained by applying various differential operators to the potential functions (7-4). The simplest of these is obtained by differentiating u_1^3/U and w_1^3/U with respect to x:

$$D_x \frac{u_1^3}{U} \equiv a \frac{\partial}{\partial x} \frac{u_1^3}{U} = - \frac{x/a}{\sqrt{1 - (x^2/a^2) - (y^2/b^2)}} \qquad (7\text{-}22)$$

and

$$D_x \frac{w_1^3}{U} \equiv a \frac{\partial}{\partial x} \frac{w_1^3}{U} = -E(k) \qquad (7\text{-}23)$$

Since the downwash is constant over the entire disk, Eq. 7-22 is a solution for a flat elliptic plate at an angle of attack. However, the flow shows infinite velocities around the trailing edge and corresponds to the noncirculatory flow depicted in part 1 of Fig. A,1a. The distribution of pressure over the sections is in fact identical with that given by formula 2 of Table A,2 for the noncirculatory two-dimensional flow over a flat plate at an angle of attack lower by the factor $1/E(k)$.

A circulatory flow over the disk is obtained by applying the operator

$$D_{ab} = 1 + a \frac{\partial}{\partial x} + b \frac{\partial}{\partial y} \qquad (7\text{-}24)$$

40 u_1/U. Geometrically, this operation amounts to a uniform infinitesimal expansion of the scale of dimensions, followed by subtraction of the original field. Application of the operator to Eq. 7-11 and 7-19 produces

$$D_{ab} \frac{u_1^3}{U} = \frac{1}{\sqrt{1 - (x^2/a^2) - (y^2/b^2)}} \qquad (7\text{-}25)$$

which is seen to be analogous to formula 1 of Table A,2, and

$$D_{ab} \frac{w_1^3}{U} = - \frac{a}{b} \frac{dQ_0(\eta)}{d\eta} = - \frac{a}{2} \left(\frac{1}{b - y} + \frac{1}{b + y} \right) \qquad (7\text{-}26)$$

Unfortunately this latter solution cannot be combined with Eq. 7-23 in such a way as to satisfy the Kutta condition all around the trailing edge. Eq. 7-23 shows a uniform (infinite) velocity around the trailing edge at all sections, while Eq. 7-25 and 7-26 show increasing velocities and lift coefficients toward the tips. The spanwise lift distribution is rectangular (i.e. $\Gamma(\eta) \sim P_0(\eta)$) and, although each section of the wing is flat, the wing is twisted to conform to the downwash induced along the center line by a pair of finite vortices emanating from the tip (see Fig. A,7c).

The determination of the flow over the flat inclined elliptic plate in such a way as to satisfy the Kutta condition at all points of the trailing edge would require the superposition of infinitely many potential functions of the type (7-25), together with the noncirculatory flow (7-22). The total lift of the wing, however, requires only the determination of the first term of such an expansion. This calculation has been carried out by Krienes for wings of several different aspect ratios and his results are reproduced in Fig. A,7d.

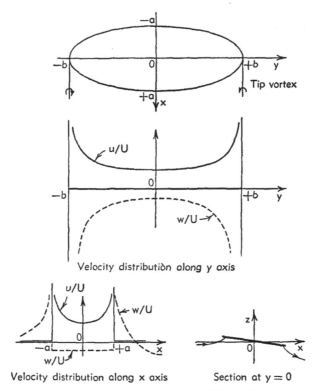

Fig. A,7c. Circulatory flow over elliptic disk.

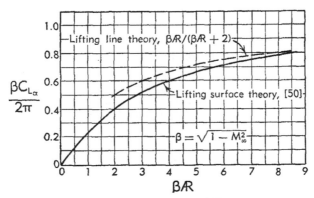

Fig. A,7d. Theoretical variations of lift curve slope
with aspect ratio for elliptic wings.

WINGS WITH UNIFORM LOADING. Another family of completely determined solutions [55,56] gives the camber and twist of straight-sided wings with uniform loading. In principle, such solutions may be found by the straightforward integration of formulas for the downwash in terms of doublet or vortex strength to be given later in this article, but in practice the integrations are not always simple.

For straight-sided planforms it is possible to write down immediately some elementary solutions which may be combined through the use of the superposition principle of linearized theory to give the desired wing outline. The elementary solutions are found by the subsonic analogue of the conical flow theory developed by Busemann [57] in connection with supersonic wing problems. Its mathematical basis is the general solution of Laplace's equation of zero degree first given by Donkin (see [34, p. 357]), namely

$$F\left(\frac{y + iz}{x + \sqrt{x^2 + y^2 + z^2}}\right) + G\left(\frac{y - iz}{x + \sqrt{x^2 + y^2 + z^2}}\right) \quad (7\text{-}27)$$

where F and G are arbitrary analytic functions. The variable

$$\epsilon = \frac{y + iz}{x + \sqrt{x^2 + y^2 + z^2}} \quad (7\text{-}28)$$

and its conjugate are expressible as functions of the ratios y/x and z/x, so that the functions $F(\epsilon)$ and $G(\bar{\epsilon})$ may be used to describe a flow field of infinite extent in which conditions are constant along rays $y/x = \text{const}$ and $z/x = \text{const}$, extending in every direction from the origin. Such fields are known as conical flow fields, and the conical flow method consists in the superposition of a number of such fields, oriented as needed to satisfy the boundary conditions of the problem. (This method will receive further application in Art. 9 and 13.)

Consider first a uniformly loaded lifting surface in the form of an infinite triangle or sector. If the origin of the coordinate system is placed at the apex of the sector, then the two boundaries of the sector may be considered rays of a conical field with apex at the origin. Let ϵ_1 and ϵ_2 ($\epsilon_1 < \epsilon_2$) be the particular values of ϵ corresponding to the two edges of the sector. Then the boundary conditions on $u(\epsilon)$ are

(i) $u(\epsilon) = \pm u_0 \text{ (const) when } z = \pm 0 \text{ and } \epsilon_1 < \epsilon < \epsilon_2$

(ii) $u(\epsilon) = 0 \text{ when } \epsilon < \epsilon_1 \text{ and } \epsilon > \epsilon_2 \text{ } (z = 0)$ (7-29)

By reference to Table A,2 (formula 8) it is found that the imaginary part of the Legendre function Q_0 shows the required variation, from which may be obtained

$$u = -\text{R.P.} \frac{iu_0}{\pi} \ln \frac{\epsilon - \epsilon_2}{\epsilon - \epsilon_1} \quad (7\text{-}30)$$

In order to find the corresponding vertical velocity we write the condition for irrotationality

$$\frac{\partial w}{\partial x} = \frac{\partial u}{\partial z} \tag{7-31}$$

or

$$\frac{dw}{d\epsilon} \cdot \frac{\partial \epsilon}{\partial x} = \frac{du}{d\epsilon} \cdot \frac{\partial \epsilon}{\partial z} \tag{7-32}$$

From this we find

$$dw = -\frac{i}{2}\left(\epsilon + \frac{1}{\epsilon}\right) du \tag{7-33}$$

so that

$$w = -\frac{i}{2} \int \left(\epsilon + \frac{1}{\epsilon}\right) \frac{du}{d\epsilon}\, d\epsilon \tag{7-34}$$

Applying Eq. 7-34 to the function (7-30) for the uniformly loaded sector and setting z equal to zero, we obtain the vertical component of the velocity in the plane of the wing:

$$w_{z=0} = -\frac{u_0}{2\pi}\left[\frac{\sqrt{1+m_2^2}}{m_2} \ln\,(M_2 - T)^2 - \frac{\sqrt{1+m_1^2}}{m_1} \ln\,(T - M_1)^2 \right.$$

$$\left. + \left|\frac{1}{M_2} - \frac{1}{M_1}\right| \sinh^{-1}\frac{1}{|t|} \right] + \text{const} \tag{7-35}$$

where $t = y/x$, m_1 and m_2 are the slopes of the two boundaries of the sector, and the functions

$$T(t) = \frac{t}{1 + \sqrt{1+t^2}}, \quad M_1 = T(m_1), \quad M_2 = T(m_2) \tag{7-36}$$

are the expressions to which ϵ, ϵ_1, and ϵ_2 reduce in the $z = 0$ plane. In the superposition of two or more infinite fields to produce a wing of finite chord, the constant of integration will be cancelled.

The contour of the surface, except for an arbitrary variation $z_0(y)$ of dihedral, is obtained by an integration of the slope w/U with respect to x:

$$z = -\frac{u_0 x}{2\pi U}\left\{ (t - m_1)\frac{\sqrt{1+m_1^2}}{m_1^2} \ln\,(T - M_1)^2 \right.$$

$$+ (m_2 - t)\frac{\sqrt{1+m_2^2}}{m_2^2} \ln\,(M_2 - T)^2$$

$$+ \left[\left(\frac{\sqrt{1+m_2^2}}{m_2} - \frac{\sqrt{1+m_1^2}}{m_1}\right)t + \left(\frac{1}{M_2} - \frac{1}{M_1}\right)\right]\sinh^{-1}\frac{1}{|t|}$$

$$\left. + \left(\frac{1}{m_1} - \frac{1}{m_2}\right)\sqrt{1+t^2} + \text{const} \right\} + z_0(y) \tag{7-37}$$

In the simplest case, z_0 will be proportional to y and the surface will be a portion of a very shallow cone about the origin, since z/x is then a function of y/x.

The process by which wings of finite dimensions are derived is illustrated in Fig. A,7e. Starting with an infinite sector having its apex at the foremost corner of the wing planform, we superpose additional sectors of positive and negative lift in such a way as to satisfy the condition of zero lift over all areas of the plane *outside* the boundary of the planform. Since solutions satisfying the conditions (7-29) do not introduce any lift outside their own uniformly loaded sectors, the superposition of such

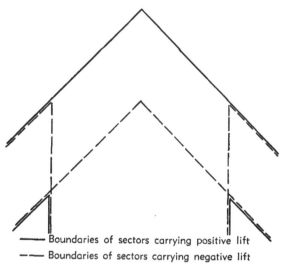

——— Boundaries of sectors carrying positive lift
— — Boundaries of sectors carrying negative lift

Fig. A,7e.   Construction of a constant chord sweptback
wing by the superposition of conical areas.

conical fields does not interfere with the condition of uniform loading on the wing.

It is found that this process requires one sector at each corner of the planform boundary. The shape of the wing $z(x, y)$ is then obtained by adding the ordinates of the stream surfaces given by Eq. 7-37, with the appropriate values of m_1 and m_2 for each superposition, and the origin displaced to the wing tips or trailing edge, as the case may be. If the stream surface associated with each sector is a conical figure, the resultant shape of the wing surface can be obtained most readily by laying out the elements (rays) of each conical surface and adding the ordinates at the points of intersection of the rays from the various fields. The dihedral may be adjusted afterwards to give a more practical shape.

The chief difficulty in the calculation of surfaces of constant loading arises in a failure of the linearized boundary condition along the root section of the triangular or sweptback wing. It will be noted on examining the form of the elementary solution for a sector of uniform loading that the downwash shows a logarithmic infinity along the x axis. The stream-lines in the x, z plane therefore cross the x axis with a vertical slope. Since the integration of w to obtain the surface shape was performed, not along a streamline, but in the $z = 0$ plane, the result, along the x axis, is an apparently infinite displacement of the surface. It can be shown that for any small value of the lift coefficient the streamline passing through the

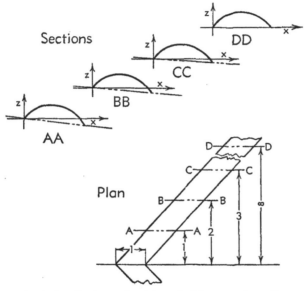

Fig. A,7f.   Camber and twist required to support a uniform
distribution of lift; $\Lambda = 45°$, $M_\infty = 0$.

origin is actually displaced only a small angle below the x axis. The varia-tion in magnitude of this angle with lift coefficient is not even approxi-mately linear, however, the initial rate of change being infinite. Moreover, the streamlines show a physically unacceptable singularity in slope (along the x axis) *above* the wing surface. While this situation invalidates a precise determination of the shape of the wing surface at the root section, it is nevertheless highly localized and may probably be disregarded if the center of the wing is to be enclosed in a fuselage. The singular behavior is removed entirely if the condition of absolute uniformity of the loading is relaxed to permit a small decrease in loading at the center section.

Lampert [55,56] has worked out several examples of uniformly loaded wings. From the practical standpoint the most interesting is the shape of a uniformly loaded, constant chord wing with 45° of sweep. This result is shown in Fig. A,7f.

If additional sectors were superposed at the tips, as in Fig. A,7e, to reduce the wing to finite aspect ratio, a prohibitive increase of twist near the tip and an infinite induced drag would result. In this case, the singularity in the downwash is of a higher order than the singularity following a bend in the leading edge. This effect is avoided if the tips are pointed or raked.

ARBITRARY LOADING; VORTEX SHEET THEORY. We proceed, from the foregoing quite special cases, to a consideration of the general problem, with loading and planform arbitrary. Because it is basic to most of the numerical methods of lifting surface theory and offers the advantage of being physically intuitive, we will discuss first the vortex sheet representation of the lifting surface. The relation between the lift and the vorticity is derived from the definition of the circulation as the line integral

$$\Gamma = \int_C \Delta \mathbf{q} \cdot d\mathbf{s} \qquad (7\text{-}38)$$

where

$$\Delta \mathbf{q} = \mathbf{i}u + \mathbf{j}v + \mathbf{k}w$$

represents the disturbance velocity due to the airfoil in an otherwise irrotational stream. Now let the path be a small flat rectangle lying in a plane $y = $ const and enclosing a portion of the wing surface. If the top and bottom of the rectangle are parallel to the wing (assumed in the $z = 0$ plane), Eq. 7-38 reduces to

$$\Gamma = \int_C (u\,dx + w\,dz) \qquad (7\text{-}39)$$

Since w is continuous across the wing (except in the neighborhood of the leading edge) the second term can be made to vanish by allowing the rectangle to become vanishingly thin. In the limit, the singularity at the leading edge can also be shown to contribute nothing to the integral. The horizontal velocity u will approach equal and oppositely-directed values at the surface of discontinuity, so that it is possible to write for the total circulation ahead of a point x, y

$$\Gamma(x, y) = 2 \int_{x_1(y)}^x u_{z=0}\, dx \qquad (7\text{-}40)$$

where $x = x_1(y)$ is the equation of the leading edge. The function $\Gamma(x, y)$ is related to the vorticity γ by the equation $\gamma = d\Gamma/dx$, and reduces to the more familiar span loading $\Gamma(y)$ at the trailing edge.

A familiar representation of the vortex sheet is as a planar arrangement of horseshoe vortices of infinitesimal spanwise extent dy and of infinitesimal strength, which will be in the present notation $\gamma\, dx$. Then by the Biot-Savart law the vertical component of the velocity at any point

x_0, y_0 in the plane of the wing may be found as

$$w(x_0, y_0) = \frac{1}{4\pi} \iint_S \gamma \left\{ \frac{x - x_0}{r^3} - \frac{\partial}{\partial y} \left[\frac{1}{y - y_0} \left(1 - \frac{x - x_0}{r} \right) \right] \right\} dx dy$$

in which $r = \sqrt{(x - x_0)^2 + (y - y_0)^2}$. The expression in square brackets is the downwash of a single trailing vortex. Taking the indicated derivative gives

$$w(x_0, y_0) = \frac{1}{4\pi} \iint_S \frac{\gamma(x, y)}{(y - y_0)^2} \left(1 - \frac{x - x_0}{r} \right) dx dy \qquad (7\text{-}41)$$

Eq. 7-41 cannot generally be evaluated analytically for useful distributions of γ. Numerical evaluation is complicated by the strong singularity in the integrand. An alternate formulation of the downwash suggested in [58] facilitates the integration to a considerable extent. In this approach, the vortices are of uniform infinitesimal strength and arranged along lines of equal Γ on the wing, a new vortex element being required for every change in circulation. In the wake there are no new sources of vorticity, so that the lines of equal Γ appear as the usual trailing vortices. Unless the surface is cambered to a vertical slope at the trailing edge the bound vortices of the wing must merge smoothly into the trailing vortices at the trailing edge of the wing in order to conform to the Kutta condition. Fig. A,7g shows an example of such a vortex representation.

By the introduction of polar coordinates r, ϑ centered in each calculation about the point x_0, y_0 at which the downwash is to be found, Eq. 7-41 is replaced by

$$w(x_0, y_0) = -\frac{1}{4\pi} \int_0^\infty \int_0^{2\pi} \frac{1}{r} \frac{\partial \Gamma}{\partial r} d\vartheta dr \qquad (7\text{-}42)$$

or

$$w(x_0, y_0) = -\frac{1}{2} \int_0^\infty \frac{1}{r} \frac{d\bar{\Gamma}}{dr} dr \qquad (7\text{-}43)$$

where $\bar{\Gamma}$ is the average value of Γ around the circle with center at x_0, y_0 and radius r. Eq. 7-43 is essentially the equation for the downwash at the initial point of a semi-infinite lifting line, which is readily evaluated by a numerical method presented in [58]. Thus the slope of the surface to support a given distribution of lift may be determined at an arbitrary number of points.

Since the determination of $\bar{\Gamma}$ is expected to be performed graphically (from a vortex pattern such as is shown in Fig. A,7g), it is not necessary to specify any analytic form for the loading. However, the accuracy will be considerably improved if the spanwise and chordwise variations of Γ can be fitted, in the neighborhood of each point x_0, y_0, with analytic expressions, to provide the initial variation of $\bar{\Gamma}$ with r.

Some Results from Lifting Surface Theory. Although the particular wings studied in [58] were elliptical in chord distribution, the results, interpreted in general terms, summarize most of what is presently known regarding the theoretical distribution of lift on wings of finite planform.

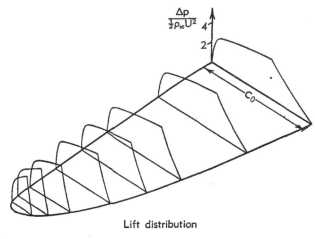

Lift distribution

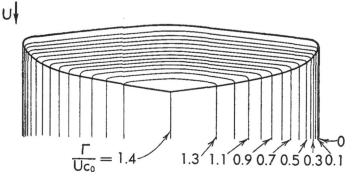

Contour lines of circulation function, or vortex pattern

Fig. A,7g. Arbitrary distribution of lifting pressure and corresponding vortex-sheet representation.

Effects of aspect ratio. In Fig. A,7h and A,7i are shown two unswept elliptical wings, $R = 6$ and 3, and the vortex patterns corresponding to the following distribution of lift: The span loading in each case is the elliptic loading predicted for an untwisted elliptic wing by the lifting line theory, and the chordwise loading at each section is that on an infinite flat plate at an angle of attack (formula 3, Table A,2).

The exact value of the downwash along the 50 per cent chord line for this loading may be obtained without calculation from the following considerations: As was pointed out earlier, in connection with the solution given by Eq. 7-22 and 7-23, the nonlifting component of the two-dimensional solution for the flat plate at an angle of attack is also a solution for the section flow on a flat elliptic disk. The downwash in the latter

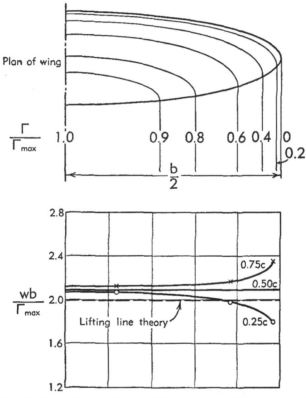

Fig. A,7h.　Vortex pattern and induced downwash for elliptical wing of aspect ratio 6, with loading based on two-dimensional flow theories.

case was shown to be E times the downwash for the same distribution of u in the two-dimensional flow, or

$$\frac{w_1}{U} = \frac{C_L}{2\pi} E \tag{7-44}$$

where E is the ratio of the semiperimeter of the ellipse to its diameter.

The lifting component of the assumed two-dimensional velocity distribution corresponds to a circulatory flow symmetrical about the 50 per cent chord line. This component cannot be related to the circulatory flow over the flat elliptic disk described by Eq. 7-25 and 7-26 because the span loading in that solution is rectangular, rather than elliptical as assumed here. However, because of the fore-and-aft symmetry of the lift distribution, the downwash along the 50 per cent chord line will be exactly half the downwash induced by the wake at infinity. For the elliptic loading, this value is

$$\frac{w_2}{U} = \frac{C_L}{\pi \mathcal{R}} \tag{7-45}$$

as given by lifting line theory. Thus the total downwash at the 50 per

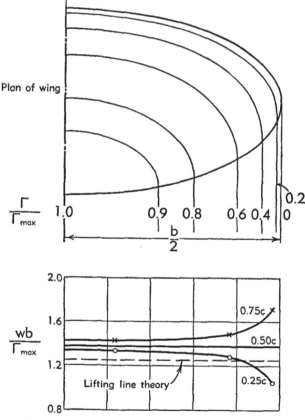

Fig. A,7i. Vortex pattern and induced downwash for elliptical wing of aspect ratio 3, with loading based on two-dimensional flow theories.

cent chord line is given by

$$\frac{w}{U} = \frac{C_L}{2\pi} \cdot \frac{E\mathcal{R} + 2}{\mathcal{R}} \tag{7-46}$$

This value and the downwash along the quarter chord and three-quarter chord lines calculated by the graphical integration are shown in the lower parts of Fig. A,7h and A,7i. The deviation of the curves from the straight lines derived by lifting line theory indicates the error in the two-dimensional theories as applied to wings of finite aspect ratio. The chief effect is seen to be an approximately symmetrical curvature of

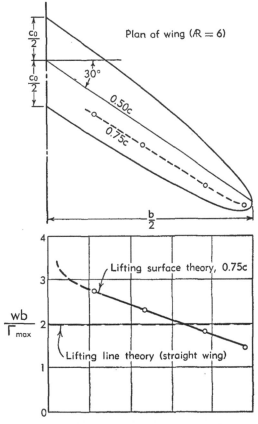

Fig. A,7j. Planform and induced downwash at three-quarter chord line of sweptback elliptical wing with loading based on two-dimensional flow theories.

the flow, in a direction requiring positive camber of the wing. The curvature is small at the center but increases sharply near the tips. The effect is of course more marked on the wing of lower aspect ratio.

Effect of sweepback. The same problem was worked for a wing with the 50 per cent chord line swept back 30°, the chord distribution remaining elliptical and the sections remaining parallel to the stream. The downwash was found only at the three-quarter chord line. The results (Fig. A,7j) indicate a linear twist for most of the wing, with more than 50 per cent reduction in angle of attack in going from a section near the root to one near the tip. Along the root section, the downwash goes to infinity, showing the assumed bending of the vortices to be physically incompatible with a continuous surface.

For the reverse problem—finding the lift distribution on a flat wing of the same planform—the vortices were rounded out at the center line to eliminate the infinite downwash of the preceding example, and the remainder of the pattern adjusted so that the downwash was approximately constant at 16 points, 8 on the quarter chord and 8 on the three-quarter chord lines.

The resulting load distribution (Fig. A,7k) shows the tendency of the load along the root section of a sweptback wing to be distributed rear-

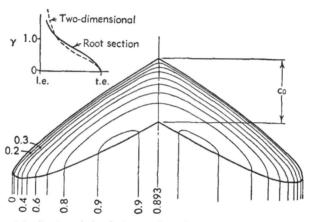

Fig. A,7k.   Distribution of circulation on flat wing of aspect ratio 6, with elliptical chord distribution swept back 30° along midchord line. Contours of the circulation function Γ, from [58], and chordwise loading at the root section.

ward compared with that predicted by two-dimensional theory. Near the tips, the opposite effect is observed, i.e. the lift tends to concentrate near the leading edge and the section aerodynamic centers are forward of the quarter chord points. The drop in lift in the center and concentration of lift at the tips are characteristic of a sweptback wing whose sweepback is not offset by washout of the angle of attack [59]. These effects are more clearly seen in Fig. A,7l, where comparison of the span loading is made with that for the unswept elliptic wing, in one case at equal lifts and in the other case at equal angles of attack. In the latter case, the introduction of sweepback results in about 14 per cent loss in total lift.

ELECTROMAGNETIC ANALOGUES.   Closely related to the foregoing investigation is the work of Swanson and Crandall [60], based on the correspondence between the velocity field of a vortex filament and the magnetic field induced by an electric current. Starting with a physical model of the vortex pattern, with the vortices replaced by wires, they find the corresponding downwash by measuring the vertical component of the magnetic field. The advantage of this method is that once the model has been constructed, the "downwash" may be measured at a very large number of points without undue expenditure of time. This advantage is partly offset by the disadvantage that numerous measurements must be made before useful results can be obtained, because of the discontinuities in field strength associated with the use of a system of discrete wires. Fig. A,7m shows a typical chordwise distribution of downwash for a model consisting of 50 wires. It should be noted also that the measurements were necessarily made at a finite height above the plane of the wing. To

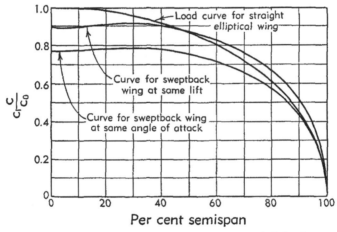

Fig. A,7l. Span load curves for wings with elliptical chord distribution, $R = 6$, showing effect of 30° of sweepback.

correct for this deviation from the analytical treatment, the measurements were repeated at several vertical stations. In spite of the oscillatory form of the curves at each level, it was found possible to obtain by fairing through the median points values which, when extrapolated to zero height, showed very smooth and consistent variations, both spanwise and chordwise.

Unpublished data obtained for the vortex configurations of Fig. A,7h and A,7i show linear chordwise variation of w, substantiating the indication of the previously obtained numerical results that the induced camber of the assumed loading is parabolic in form. Extension of the investigation to the problem of the straight elliptical wing in steady rolling motion [61] yielded a similar result, with the same sort of increase in camber toward the tip as with the angle-of-attack loading.

It may be appropriate to mention here a somewhat different electrical analogue—the electrolytic tank. Apparently successful tests have been made by Malavard [*62*] (see also IX,H,2), replacing the wing with a large number of small conducting plates, and the wake with strips of the same, in order to generate an electric potential to correspond to the velocity potential of the wing. The resistances in series with the plates are variable. The quantity measured is the current through the plate, which, being proportional to the gradient of the potential, corresponds to the downwash in the aerodynamic problem. This method is discussed further below.

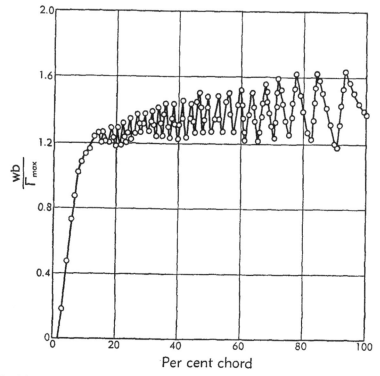

Fig. A,7m. Variation of the downwash measured one per cent of the semispan above the center section of the electromagnetic analogue of an elliptic wing, with two-dimensional flat plate loading approximated by 50 wires [*60*].

THE "DIRECT PROBLEM"—THE WING COMPLETELY SPECIFIED. In the foregoing, methods were presented for solving the so-called "indirect problem" of aerodynamics—that is, finding the shape of the body consistent with an assumed distribution of pressure. This type of problem is inherently easier to solve for the following reason: beyond the boundaries of the lifting surface the pressure, or u component of the perturbation velocity, must be zero. If the distribution of pressure is specified on the surface also, then the function u is described all over the plane and the related quantities φ and w can in principle be found by straightforward integration. Moreover, new solutions can be found by the super-position of simple solutions, as was done in deriving the uniformly loaded wing, since the component solutions introduce no new lift beyond their own surfaces.

However, the majority of the designer's problems are of the reverse type: he is required to predict the effect on the lift distribution of control-surface deflection, of angle of attack, of specified twist, and so on. In these problems, the boundary condition of tangential flow prescribes the variation of w on the wing, but elsewhere the condition $u = 0$ must still be satisfied. Thus we have a problem with mixed boundary conditions leading, in the case of a finite wing, to an integral equation for which no analytical method of solution has as yet been developed.

The previously mentioned "rheoelectric" method of Malavard (see IX,H,2) furnishes one method of solution through the adjustment of the resistances until the required variation of current (proportional to the downwash) is obtained. By this method span loadings and center-of-pressure locations have been found [*62*] for wings with various degrees of sweepback, including the effects of flaps and ailerons. Some work has also been done on the development of relaxation methods [*63,64*] and satisfactory results obtained.

While the detailed solutions obtainable by such methods would be of value as a basis for boundary layer studies, for flap-effect calculations, and for checking the accuracy of more approximate methods, the special equipment or lengthy calculations involved make the methods not en-tirely suitable for general use. Instead, the most usual procedure has been to superimpose a number of solutions to the indirect problem, in such proportions as to satisfy the conditions on the vertical component of the velocity at a finite number of points on the wing.

The methods of approximation used to reduce the computing to practical proportions generally take one of two forms: either (1) assuming the chordwise load distribution to be adequately expressed by a small number of terms of a trigonometric series, or (2) replacing the continuous vortex sheet with a finite number of discrete vortices.

Lifting surface methods. If a continuous distribution of vorticity is retained, it is generally assumed, after Birnbaum's treatment of the two-dimensional problem [*65*], to have along each chordwise section the form

$$\gamma = a_0 \cot \frac{\theta}{2} + \sum_{n=1}^{N} a_n \sin n\theta \qquad (7\text{-}47)$$

$$\theta = \cos^{-1} \frac{x}{c/2} \qquad (7\text{-}48)$$

where c is the chord of the section and x is the streamwise distance from the midpoint. The first term of this expression for γ is the trigonometric form of formula 3 of Table A,2 and corresponds to the two-dimensional solution for a flat plate at an angle of attack; the second term ($n=1$) is the solution for parabolic camber (formula 5). For a wing of finite span, the coefficients a_n are expressed as functions of the spanwise coordinate y. This approach was first used by Blenk [66] and has been the basis of most of the subsequent lifting surface work.

With the insertion of the first terms of Eq. 7-47 for γ, the chordwise integration of Eq. 7-41 may be performed analytically (in terms of elliptic integrals [67] or numerically). Blenk chose to expand the integrand in a series which is unfortunately not convergent at the wing tips or at the root of a sweptback wing. Very recently Multhopp has prepared some charts [68,69] giving numerical values for results of the first integration. With these values, the equation for w may be put into the form of the Prandtl lifting line equation and treated similarly, using Multhopp's well-established method [70] for that problem.

Like the form of the vorticity distribution, the control points at which the boundary condition of tangential flow is to be satisfied are also selected on the basis of two-dimensional theory. If, for example, the downwash is to be fixed at only one point of a given section, the three-quarter chord point is indicated as most representative for the determination of the section lift. The reason for this choice is the fact that, in two-dimensional flow, a flat plate and a wing with parabolic camber, if they coincide in slope at the three-quarter chord line, are found to have the same lift coefficient. Thus, whether the wing is uncambered or is cambered along a parabolic arc, the slope at the three-quarter chord will suffice to indicate the lift coefficient, in two-dimensional flow. Where chordwise discontinuities in slope are involved, as in the case of deflected flaps, the three-quarter chord line, of course, has no special significance, but some adjustment is made, usually in the form of the use of "equivalent slopes" for the sections, so that the three-quarter chord points can still be used.

If more than one control point per section is to be used, a formula due to Multhopp gives the location of the jth point,

$$\theta_j = \frac{2\pi j}{2p + 1}, \qquad j = 1, 2, \ldots, p \qquad (7\text{-}49)$$

where p is the total number of control points for the section.

It is generally concluded, in investigations based on the foregoing method, that one term in the series (7-47) (that is, the infinite flat plate solution) and one set of downwash points along the three-quarter chord line are sufficient for the determination of the lift and span loading of most wings. For the calculation of the aerodynamic center, at least two terms and two downwash points are required at each section.

Presumably, with more terms in the series and more control points, the surface load distribution could be worked out in greater detail. It is clear, however, that such a procedure is not really adapted to the investigation of the loading in regions of rapidly changing boundary conditions, because of the reliance on two-dimensional solutions. Moreover, it can be shown [71] that the form of loading assumed (Eq. 7-47 and 7-48) is fundamentally unsuitable for any wing containing discontinuities in the direction of the lines of constant per cent chord. In such cases (i.e. swept wings), the use of the per cent chord as a variable necessarily introduces inadmissible infinities in the vertical velocity, due to the assumed angularity of the vortices, with the result that the boundary conditions cannot be satisfied by any finite number of solutions.

Various methods of dealing with the failure of the two-dimensional approach at the root of the swept wing have been proposed. Multhopp [68] relies on the smoothing effect of the interpolation functions used in performing the spanwise integrations. Küchemann [72] has undertaken a more basic consideration of the problem, attempting to isolate the effect of the bend in the wing outline. His results, though interesting, do not lead to a complete lifting surface theory and appear to be of limited value because of the numerous assumptions introduced. Garner has attempted to solve the problem by a more suitable choice of variable, one which while conforming to the planform boundaries has, as its intermediate contours, lines of continuous curvature. Unfortunately, other mathematical difficulties (see the brief discussion in [73]) prevent a satisfactory resolution of the center-section problem.

Thus there remains as yet unsolved the question of the precise theoretical form of the lift distribution at the root of a swept wing.[8] Detailed solutions for the neighborhood of the tips, or of deflected flaps, are likewise beyond the reach of facile computing routines. Such failures of the theory need not, however, have any practical significance in determining such over-all characteristics of the wing as lift curve slope, span loading, and induced drag, since the effect of details of the loading at any one station dies away very rapidly. The chief difficulty appears when one of a small number of "control points" is located on the center section of a swept wing, since the inaccuracies introduced there then affect the entire solution.

Methods using discrete vortices. It is logical to attempt to simplify the computations by the substitution of a number of discrete vortices for the continuous distribution of circulation of the vortex sheet. The spanwise continuity of the loading is generally retained, but the lift is gathered chordwise into a finite number of lifting lines, each extending over the span of the wing. The trailing vortices leave each line in the same way as in the Prandtl lifting line model. When the wings are tapered or swept,

---

[8] An interesting comparison of the span loadings calculated for an infinite swept wing by various methods is given in [74].

the lifting lines are made to pass through points of constant per cent chord.

The formula for the downwash induced by a single lifting line with variable circulation $\Gamma(y)$, swept back through an angle Λ at the center, is

$$w(x_0, y_0) = \frac{1}{4\pi} \left[\int_{-b/2}^{b/2} \frac{\Gamma'(y)}{y - y_0} \left(1 + \frac{x_0 - |y| \tan \Lambda}{r} \right) dy \right.$$

$$\left. - (x_0 + y_0 \tan \Lambda) \int_{-b/2}^{0} \frac{\Gamma(y)}{r^3} dy - (x_0 - y_0 \tan \Lambda) \int_{0}^{b/2} \frac{\Gamma(y)}{r^3} dy \right] \quad (7\text{-}50)$$

with $r = [(x_0 - |y| \tan \Lambda)^2 + (y_0 - y)^2]^{\frac{1}{2}}$ and $y_0 \geqq 0$.

In methods employing discrete vortices, two-dimensional theory is used to determine the most representative location of the vortices as well as of the control points. If only one vortex line is used, it is placed along the center-of-pressure line in two-dimensional flow, which for a flat plate at an angle of attack is the quarter chord line.

The downwash then varies inversely as the distance behind the quarter chord line; at the position of the three-quarter chord line it just equals in magnitude the vertical component of the flow tangential to the flat plate having the same circulation. Conversely, if the condition of tangential flow is satisfied at the three-quarter chord line, the strength of the concentrated vortex will indicate the lift on the wing due to angle of attack.

The foregoing approach to the lifting surface problem—a first correction to the Prandtl lifting line theory—was initially suggested by Pistolesi [75]. Mutterperl [76] and Weissinger [77] developed the procedure further and also extended the formulas to cover sweptback wings. A side investigation by Weissinger indicated that the span loadings obtained might be expected to be in good agreement with the results of the vortex sheet calculations when the latter included only one term in the series (7-47) for the chordwise distribution of lift. Using Weissinger's formulas, De Young and Harper [78] have developed a rapid computing procedure for the span loading of wings with quarter chord lines composed of two straight-line segments and have prepared charts from which the span loading of straight-tapered wings may be obtained directly. A comparison of their results with the limited results of lifting surface methods now available[9] shows the angle-of-attack span loadings obtained by the Weissinger method to be in fact almost indistinguishable from those obtained with any of the more elaborate methods.[10]

---

[9] Covering 0 to 45° of sweep, aspect ratios from 3 to 8, and taper ratios from 0 to 1.

[10] An approximate integration formula for obtaining the total lift appears to be less reliable, since it does not gives results approaching the correct limit for infinite aspect ratio when applied to sweptback wings. However, this and other difficulties at high aspect ratios may be remedied by increasing the number of control points over the seven used in [78]. (See, for example, [79].)

Concentration of the lift in one vortex line makes it impossible, of course, to determine anything about the chordwise distribution of the load. For more detailed information, two or more vortex lines are required. Various arrangements of the vortices have been proposed. One, applied by Weighardt [80] with considerable success to rectangular wings of low aspect ratio, appears to have the advantage of simplicity of concept and application. Schlichting [81] and Scholz [82] have shown that it has also a certain physical justification in two-dimensional flow. The essence of the method is as follows:

The wing is first divided into N strips of equal depth extending from tip to tip. (In the case of a tapered wing, the strips would taper in the same ratio.) A lifting line is placed along the quarter chord line of each strip and the condition of tangential flow satisfied at M points on each three-quarter chord line. Thus, $M \cdot N$ equations are obtained for the circulation on each lifting line, which may then be expressed in the form of M terms of a Fourier series. For the physical justification of this strip method, we again turn to the two-dimensional solutions. It can be shown [83] that for any N, the sum of the circulations obtained for each of the vortices by satisfying the condition $\alpha =$ const at each of the three-quarter chord lines on an infinite wing is exactly equal to the circulation of the flat plate at that angle of attack, and the pitching moment is also reproduced. Moreover, the lift of a wing with parabolic camber is duplicated by the same system of vortices if the corresponding boundary conditions are inserted. (It is possible only to approximate the moment of the cambered wing, and this requires three or four strips.)

The discrete vortices are introduced only for the purpose of computing the downwash. The aerodynamic characteristics of the wing are computed from a continuous loading in the form of Eq. 7-47, related to the loading of the discrete vortices by equating section lifts, moments, etc. on the assumption of two-dimensional flow at each section. Of course, this method of obtaining correspondence between the lifting lines and the vortex sheet loses its validity near the corners of the wing, where the flow differs sharply from the two-dimensional.

This method has been applied by Scholz to rectangular wings, both flat and cambered. The results for flat wings at an angle of attack (Fig. A,7n) show a tendency of the span loadings toward the elliptical as the aspect ratio is reduced, and a tendency for the lift to be concentrated at the leading edge. However, the previously noted inaccuracy of the solution near the tips causes the results for the lowest aspect ratio to be only qualitatively correct, as may be seen by the plots (Fig. A,7n, bottom) of aerodynamic center location. The use of more vortices for low aspect ratios (here $N = 3$) would be expected to show a more marked forward movement of the aerodynamic center near the tips for the case $\mathcal{R} = 1$, in conformity with the other curves. In the exact potential-flow solution the

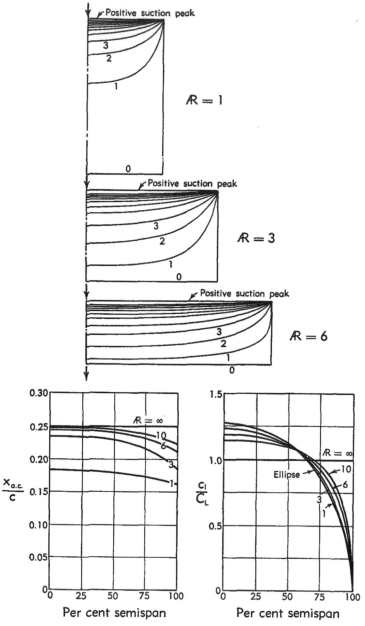

Fig. A,7n. Effect of aspect ratio on loading of flat rectangular wings, calculated by the strip method of [82]. Top, contour lines of load distribution; bottom left, aerodynamic center location aft of leading edge; bottom right, span loading.

aerodynamic center of the tip section on rectangular wings would be at the leading edge.

The contour lines of the load distribution for a rectangular wing of aspect ratio 6 with parabolic camber is shown in Fig. A,7o. It is seen that in this case, even at so large an aspect ratio, the departure from two-dimensional conditions affects the entire flow, so that the method of solution is itself invalidated.

The strip method just described has been applied to sweptback wings by Schlichting [81]. Other methods of distributing the lifting lines and the points for computing the downwash are investigated in a paper by Holme [84]. Using two lifting lines and fifteen control sections (effectively thirty control points), Holme obtains, for a wing with 40° sweepback and aspect

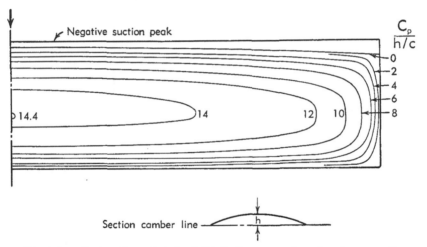

Fig. A,7o. Contour lines of the load distribution on a rectangular wing, $\mathbb{R} = 6$, with parabolic camber, calculated by the strip method of [82].

ratio 4.5, both span loading and aerodynamic center locations in remarkable agreement with experiment (Fig. A,7p). The reduction of lift and rearward movement of the section aerodynamic center at the wing root, and the opposite tendency near the tips—indicated earlier in this article as the effects of sweepback—are here evident in the experimental data.

The foregoing procedures are essentially extensions of the lifting line theory, with the formulas modified to admit one bend in the lifting lines. It is apparent that the range of planforms to which this representation can be applied is somewhat limited, since the introduction of additional corners, or curves, would overcomplicate the formulas. Thwaites [85] has suggested the use of broken-line vortices adaptable to any planform, with the compensating simplification of a quadratic approximation to the variation of vorticity along each segment. By adjusting the constants, the discontinuities in the direction of the vortices and in the magnitude of the loading can be made to counteract each other so that no singularities occur in the downwash. This procedure would appear to avoid many of the pitfalls of other discrete-vortex methods but does not seem to have been developed to any extent.

Vortex lattice method. An arrangement designed to permit greater flexibility in the range of planforms treated has been developed by Falkner

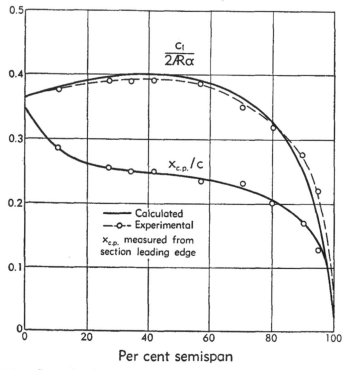

Fig. A,7p. Comparison between calculated and experimental lift distributions and center-of-pressure positions for sweptback untapered wing; $\mathcal{R}$ = 4.5, Λ = 40° [84].

[86,87,88]. It consists of a large number (of the order of 100) of horseshoe vortices, each extending over only a small fraction of the span, placed so as to form a rectangular lattice, of which Fig. A,7q shows a typical example. By this method the loading along the span as well as along the chord is made stepwise discontinuous. Use is made of extensive tables of the downwash field of a horseshoe vortex computed by the National Physics Laboratory [89]. The procedure is similar in other respects to the previously discussed discrete-vortex methods.

In any method employing discrete vortices, the relative location of vortices and control points is of critical importance, since the downwash tends toward infinity in the neighborhood of each vortex. When only a single vortex line is used and the control points are located on the three-quarter chord line, as in the preceding methods, their distance from the vortex is large enough so that the variation of the downwash with location is slow. When the number of vortices is large, the downwash varies from point to point in a manner similar to that shown in Fig. A,7m. It is not surprising, then, that in Falkner's method the solution appears to depend on the vortex pattern and on the control points employed.

It should be noted that Multhopp's lifting surface method, which is equally flexible as regards the planforms that can be treated, is not subject

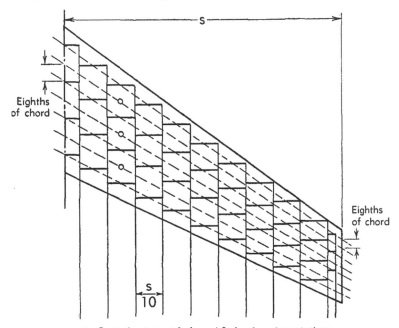

○ Control points at $\frac{1}{4}$, $\frac{1}{2}$, and $\frac{3}{4}$ chord, various stations

Fig. A,7q. Typical vortex pattern and location of control points for Falkner's vortex-lattice method.

to the objections raised above and, with the publication of the necessary tables [69], becomes equally usable. Similar conclusions have been reached by Garner [73] in a recently published evaluation of the more widely used lifting surface methods.

EFFECT OF PLANFORM ON LIFT DUE TO ANGLE OF ATTACK. In Fig. A,7r are shown the effects of taper and sweep on the theoretical span loading of wings of aspect ratio 4. The curves are derived from Falkner's results and the charts of [73], which are consistent for the cases shown. The values of $C_{L\alpha}$ are those given by Falkner. It can be seen that at low angles of attack the effects of taper and sweepback tend to counteract each other, both with respect to the total lift and its distribution along the span. However, since the lift per unit chord length is greater near the tip of the tapered wing, the undesirable stalling characteristics of a sweptback wing are worsened by the use of taper.

In Fig. A,7s is shown the variation of lift curve slope C_{L_α} with aspect ratio according to various lifting surface theories, for untapered wings with 0, 30, and 45 degrees of sweepback. Included for comparison are the results of the Prandtl lifting line theory, which is of course applicable only to straight wings, and the low aspect ratio theory to be discussed later in

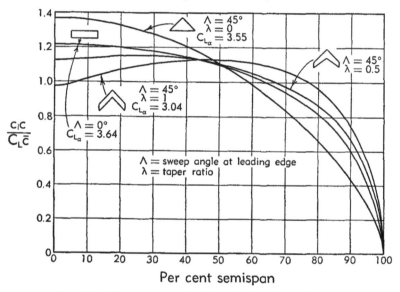

Fig. A,7r. Theoretical span loadings for four wings of $Æ = 4$.

this article. The latter theory gives a formula for C_{L_α} dependent only on the aspect ratio.

Since any of the foregoing methods may be applied to problems in compressible flow by the use of the Prandtl correction (Art. 5), the parameters of Fig. A,7s are written to include the effects of compressibility. The angle of sweepback Λ is modified by the transformation

$$\tan \Lambda_{\text{comp}} = \frac{\tan \Lambda}{\sqrt{1 - M_\infty^2}} \tag{7-51}$$

The lift curve slope and aspect ratio are simply multiplied by $\sqrt{1 - M_\infty^2}$.

The curve for zero sweep is drawn through values taken from [66,81, 82,84,87]. While the results of these various lifting surface methods are in excellent agreement as regards the simple rectangular wing, the indication with respect to the lift of sweptback wings is not quite so clear. Curves from [76] and [87] for $\Lambda_{comp} = 30°$ and $45°$ are in good agreement, but Schlichting's curve for the latter case is somewhat higher and Weissinger's (from [78]) considerably lower. It has already been mentioned that Weissinger's results fail to approach the correct asymptotes (Eq. 4-8) for infinite span, and are therefore not reliable for higher aspect ratios. However, the asymptotic slope at zero aspect ratio is correct. Mutterperl's formula approaches the correct values at either extreme. No further values from lifting surface theory are available for the particular case of the untapered wing. Fig. A,7s includes, however, the results given

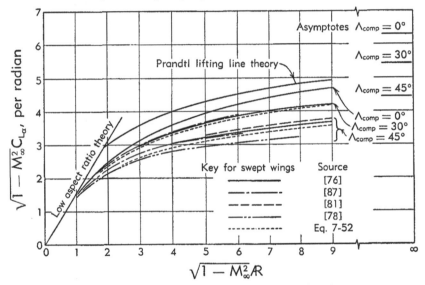

Fig. A,7s. Theoretical values of the lift curve slope for untapered wings with various degrees of sweepback.

by an easily applied empirical formula which reproduces the theoretical curve for the rectangular wings down to aspect ratios close to one, as well as the values for several tapered sweptback wings to which the lifting surface theories have been applied. This formula is

$$C_{L\alpha} \cong \frac{2\pi R}{pR + 2} \tag{7-52}$$

where p is the ratio of the semiperimeter of the wing to its span, and results from the generalization of the first order correction to the lifting line theory derived in [54] for elliptical wings (cf. also Eq. 7-46). Applied to sweptback untapered wings, Eq. 7-52 gives values somewhat lower than the lifting surface values, but perhaps close enough for engineering requirements.

Experimental values of the lift curve slope of untapered wings, obtained from a number of sources, are plotted in Fig. A,7t and compared with the theoretical curves from [76], extended below $R = 3$ by the use of Falkner's results. In all the theoretical work discussed, 2π has been tacitly assumed as the value of C_{L_α} in two-dimensional flow, whereas it is well known that higher or lower values have been obtained in wind tunnel tests, largely depending on the trailing edge angle of the section. The data presented in Fig. A,7t are almost wholly for conventional 9 or 12

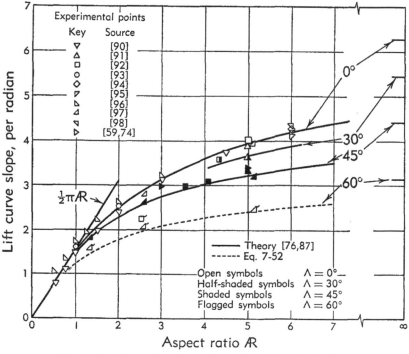

Fig. A,7t.   Comparison of experimental and theoretical values of the lift curve slope for untapered wings.

per cent thick airfoils and therefore show very little scatter. It is seen that the points are in very good agreement with the theoretical curves. On this basis the approximate formula (7-52) may also be considered a practical means of obtaining values of the lift curve slope for other angles of sweepback and for tapered wings of moderate-to-high aspect ratio.

Fig. A,7u shows a number of experimental results for triangular wings as well as the theoretical curve calculated for such wings in [78] on the basis of Weissinger's method. This curve is seen to represent the average variation of the experimental values quite well. Two results from Falkner's work indicate a slightly higher curve. Application of Eq. 7-52 to triangular wings gives a curve which, although approaching zero along the correct tangent, appears to be perhaps ten per cent too high at higher aspect ratios. The subject of triangular wings will be treated in more detail in the following article.

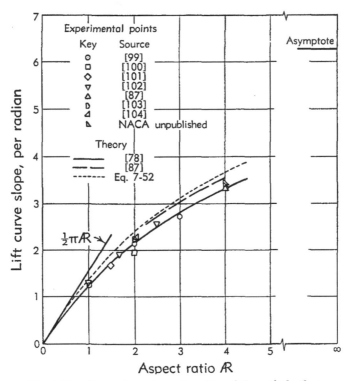

Fig. A,7u. Comparison of experimental and theoretical values of the lift curve slope for triangular wings.

A,8. Lifting Surfaces of Slender Planform. In cases of extreme sweep or very low aspect ratio the assumption of two-dimensional flow, used in the development of the theory of wings of high aspect ratio, can again be employed [105]. In this case, however, the planes of the two-dimensional motion will be perpendicular to the flight direction and the origin of the lift appears in a somewhat different aspect than in the case of the wing of large span and narrow chord. The formulas obtained resemble in some respects those derived by Munk [106] and Tsien [107] for the lift of an elongated body of revolution.

Theory for wings of low aspect ratio. With the idealization of two-dimensional flow, the treatment of the low aspect ratio wing becomes exceedingly simple. The case lending itself most readily to the analysis is that of the long flat triangular airfoil, traveling point-foremost at a small angle of attack. Viewed from a reference system at rest in the undisturbed fluid, the flow pattern in a plane cutting the airfoil at a distance x from the nose is the familiar two-dimensional flow caused by a flat plate having the normal velocity $U\alpha$ (see Fig. A,8a). Observed in this plane, the width of the plate and hence the scale of the flow pattern continually increase as the airfoil progresses through the plane. This increase in the

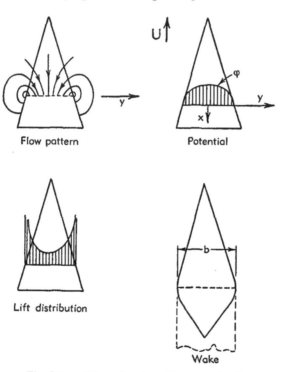

Fig. A,8a.   Triangular wing of low aspect ratio.

scale of the flow pattern requires a local lift force l equal to the downward velocity $U\alpha$ times the local rate of increase of the additional apparent mass m', or

$$l = U\alpha \frac{dm'}{dt} = U^2\alpha \frac{dm'}{dx}$$

since $U = dx/dt$.

By a well-known formula from two-dimensional flow theory,

$$m' = \pi y_1^2 \rho_\infty$$

where $y_1(x)$ is the y coordinate of the leading edge and $2y_1$ is the local width of the plate. Hence

$$\frac{dm'}{dx} = 2\pi\rho_\infty y_1 \frac{dy_1}{dx}$$

and the lift l per unit length will be given by the expression

$$l = 4\pi\alpha \frac{1}{2}\rho U^2 y_1 \frac{dy_1}{dx}$$

Dividing by $\frac{1}{2}\rho_\infty U^2$ and by the span $2y_1$ gives the local lift coefficient

$$c_l = 2\pi\alpha \frac{dy_1}{dx} \tag{8-1}$$

When this flow is considered in more detail, it is found from the two-dimensional theory that the surface potential φ is distributed spanwise according to the ordinates of an ellipse, i.e.

$$\varphi = \pm U\alpha \sqrt{y_1^2 - y^2} \tag{8-2}$$

(see Fig. A,8a). An instant later, in the same plane, the ordinates are those of a slightly larger ellipse, corresponding to an increase of φ. The local pressure difference is given by the local rate of increase of φ,

$$p_{\text{lower}} - p_{\text{upper}} = \Delta p = 2\rho_\infty \frac{\partial\varphi}{\partial t}$$

$$= 2\rho_\infty U \frac{\partial\varphi}{\partial x}$$

$$= 2\rho_\infty U \frac{\partial\varphi}{\partial y_1} \frac{dy_1}{dx} \tag{8-3}$$

where $\partial\varphi/\partial y_1$ is a function of y. Differentiation of φ yields the equation

$$\Delta p = 2\rho_\infty U^2\alpha \frac{y_1}{\sqrt{y_1^2 - y^2}} \frac{dy_1}{dx} \tag{8-4}$$

The pressure distribution thus shows an infinite peak along the sloping sides of the airfoil similar to the pressure peak at the leading edge of a conventional airfoil. The distribution along radial lines passing through the vertex of the triangle (lines of constant y/y_1) is uniform, however, and the center of pressure coincides with the center of area. The maintenance of lift right up to the trailing edge is an effect associated solely with the limiting case of zero width and will not exist in the case of a wing of finite aspect ratio.

Eq. 8-1 and 8-4, although derived for triangular wings, actually apply to more general planforms, since dy_1/dx is not necessarily a constant, and show that the development of lift by a long slender airfoil depends on an expansion of the sections in a downstream direction. Hence a part of the surface having parallel sides would develop no lift. Furthermore, a decreasing width would, according to Eq. 8-4, require negative lift with infinite negative pressure peaks along the edges of the narrower sections. In the actual flow, however, the edge behind the maximum cross section will lie in the viscous or turbulent wake formed over the surface ahead; for this reason it will be assumed that the infinite pressure difference indicated by Eq. 8-4 cannot be developed across these edges. It is this assumption, corresponding to the Kutta condition, that gives the plate the properties of an airfoil as distinct from another type of body such as a body of revolution.

With the aid of the Kutta condition, it may easily be shown that sections of the airfoil behind the section of greatest width develop no lift. A potential flow satisfying both the boundary condition and the Kutta condition may be obtained by the introduction of a free surface of discontinuity behind the widest section. This surface of discontinuity (Fig. A,8a) is composed of parallel vortices extending downstream from the widest section of the airfoil as prolongations of the vortices representing the discontinuity of potential over the forward part of the airfoil. Such a sheet, although possibly wider than the downstream sections of the airfoil, still satisfies their boundary condition, since the lateral arrangement of the vortices is such as to give uniform downward velocity equal to $U\alpha$ over the entire width of the sheet, including the rearward portion of the airfoil. The Kutta condition will be satisfied within the region covered by the sheet of parallel vortices regardless of the shape of the trailing edge. Since the pressure difference across the airfoil is proportional to $\partial\varphi/\partial x$, and since this gradient disappears as soon as the vortices become parallel to the stream, no lift is developed on the rearward sections.

Integration of the lifting pressures in a chordwise direction from the leading edge downstream to the widest section will give the span load distribution and the induced drag. Since, from Eq. 8-3,

$$\int \Delta p\,dx = 2\rho_\infty U\varphi$$

the span load distribution is

$$\frac{\partial L}{\partial y} = 2\rho_\infty U^2\alpha \sqrt{\left(\frac{b}{2}\right)^2 - y^2} \tag{8-5}$$

where b is the span of the wing. Hence $\partial L/\partial y$ is elliptical, and is furthermore independent of the planform in these simple cases.

With the elliptical span load the induced drag is equal to its minimum value

$$D_\mathrm{i} = \frac{L^2}{\pi q b^2} \tag{8-6}$$

A second integration of $(\partial L/\partial y)dy$ across the widest section gives the total lift, which is

$$L = \frac{\pi}{4} \rho_\infty U^2 \alpha b^2 \tag{8-7}$$

The lift of the slender airfoil therefore depends only on the width and not on the area.

Dividing by $\frac{1}{2}\rho_\infty U^2 S$, we obtain the lift coefficient

$$C_L = \frac{\pi}{2} \mathcal{R}\alpha, \qquad \mathcal{R} = \frac{b^2}{S} \tag{8-8}$$

and, from Eq. 8-6, the induced-drag coefficient

$$C_{D_i} = \frac{C_L^2}{\pi \mathcal{R}} = C_L \frac{\alpha}{2} \tag{8-9}$$

From Eq. 8-9 it appears that the resultant force lies halfway between the normal to the surface and the normal to the air stream.

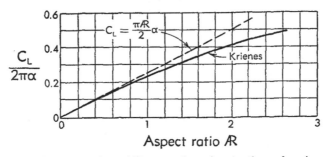

Fig. A,8b. Comparison of lift curve slope given by theory for wings of low aspect ratio with results given by Krienes [50].

If the rate of increase of the width becomes too great, the flow cannot be expected to remain two-dimensional. It is seen that in the case of a rectangular planform the simplified formula (8-1) gives an infinite concentration of lift at the leading edge and no lift elsewhere, whereas a more accurate theory would show some distribution of the lift rearward. However, it can be shown by examination of the three-dimensional (non-lifting) potential flow around an elliptic disk (Eq. 7-22 and 7-23), that the two-dimensional theory gives a good approximation in the case of an elliptical leading edge, which indicates that the theory is applicable over a large range of nose shapes. In Fig. A,8b is shown a comparison of the lift calculated by the present theory for elliptical wings of low aspect ratio with the results of the more accurate three-dimensional potential flow calculations of Krienes (Art.7). It is seen that the simple theory approximates the exact value very closely up to an aspect ratio of 1. Comparison with the lifting surface values of the lift curve slope for rectangular and sweptback wings (Fig. A,7s) as well as experimental data (Fig. A,7t) suggests the same range of validity for untapered wings.

Extensions of low aspect ratio theory. The essentially elliptic form of the span loading on low aspect ratio wings was surmised by Weighardt [80] on empirical grounds and used to reduce the lifting surface integral equation (7-41) to a one-dimensional one, which he then solved numerically for rectangular wings by means of an expansion chordwise in a Birnbaum series, satisfying the downwash condition at points along the axis of the wing.

A development which shows the transition from the theory of lifting surfaces of high aspect ratio to those of low aspect ratio has been given by Lawrence [108], and results in a more rigorous proof that the ellipse is the limiting form of the span loading on a flat rectangular plate of infinite depth. Applying the slender wing relations between w and φ derived for the rectangular wing (not necessarily flat) to simplify the lifting surface integral equation (7-41), Lawrence is able to write a first order integral equation in the integrated lift per unit chord, which is then solved numerically through expansion in series. In effect, the procedure involves equating a weighted average of the downwash at each spanwise section to the similarly weighted average of the local angle of attack. The resulting lift curve slopes for rectangular wings coincide with the lifting surface values shown in Fig. A,7s over the entire range of aspect ratios for which the calculations were made ($R \leqq 4$), whereas Weighardt's similar but more elementary approach produced results which cannot be considered accurate for aspect ratios greater than 2. For triangular wings, Lawrence's method gives values of the lift almost identical with Weissinger's (Fig. A,7u).

The problem of extending the theory to include wings of wider planform has been treated more recently in a paper by Sears and Adams [109]. By appropriate transformations of the original three-dimensional flow equation they were able to obtain formulas in terms of an expansion in powers of a width parameter. Studies of the few examples in which the three-dimensional flow is known exactly indicate that such an approach should yield accurate results.

Slender sweptback wings. The consideration of wings with sweptback trailing edges introduces additional difficulties which even a return to the two-dimensional approach fails to eliminate if the trailing edge cutout extends forward of the tip region, i.e. into the lifting part of the wing (see Fig. A,8c). This problem has been studied by Heaslet and Lomax [110] and Robinson [111], and subsequently by Mangler [112], Legendre [113] and Mirels [114].[11]

---

[11] Also, in a very recent paper [115], Eichelbrenner has indicated how the method of Adams and Sears [109] can be extended to this problem.

In cross stream sections the portion of the wing between sections AA and BB in Fig. A,8c appears as two parallel flat plates, and the form of the complex velocity function can be written down from a consideration of its singularities, except that the magnitude of the circulatory term will depend on the vorticity in the wake, which in turn depends on the spanwise variation of lift at upstream sections. The expression of this dependency on upstream conditions takes the form of an integral equation. Except for a special case, which will be discussed later, this equation has been solved only numerically.

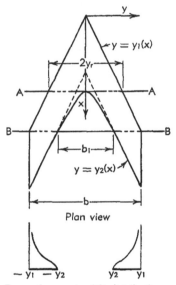

Plan view

Form of spanwise lift distribution

Fig. A,8c.   Slender sweptback wing.

We note that the problem involves four edges, at each of which the pressure either is infinite as the reciprocal of the square root, or vanishes as the square root of the distance, so that the complete solution must involve elliptic functions. From [112] and [114] we find, for a wing with arbitrary leading and trailing edges, the coefficient of lifting pressure for the flat wing at an angle of attack α to be

$$\frac{\Delta p}{\frac{1}{2}\rho_\infty U^2} = 4\alpha S(x) \frac{dy_1}{dx}\left[Z(\psi, k) + \frac{y}{y_1}\sqrt{\frac{y^2 - y_2^2}{y_4^2 - y^2}}\right] \tag{8-10}$$

where Z is the Jacobian zeta function of argument

$$\psi = \sin^{-1}\sqrt{\frac{y_1^2 - y^2}{y_1^2 - y_2^2}}$$

and modulus

$$k = \sqrt{1 - \left(\frac{y_2}{y_1}\right)^2}$$

and where $S(x)$ is a function to be determined from the integral equation.

Mirels has calculated values of $S(x)$ for untapered sweptback wings, and Mangler, for both tapered and untapered. Since the results of slender wing theory are of particular interest in their application to sonic speed, as explained in a later section, these results will be discussed further in connection with supersonic lift (Art. 14).

An example that lends itself to solution in closed form is provided by the condition that no vortices exist in the V-shaped region between the two wing panels. Ordinarily this condition would require a certain variation of the camber and twist of the wing surface. However, it is found

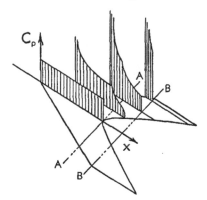

Fig. A,8d. Lift distribution on wing of Fig. A,8c (from [*110*]).

[*110*] that a flat untwisted wing meets the condition, provided the trailing edge is given a particular shape. If the leading edge is straight, the trailing edge shape required is given by the equation

$$\frac{y_2}{y_r} = \frac{y_2/y_1}{E - (y_2/y_1)^2 K} \tag{8-11}$$

where E and K are the complete elliptic integrals with the modulus $\sqrt{1 - (y_2/y_1)^2}$ and y_r is the semispan of the section AA through the trailing edge of the root chord. With the substitution of $y_1 \sim x$, Eq. 8-11 gives the equation of the trailing edge in parametric form. The shape is the one used for illustration in Fig. A,8c. At large aspect ratios, the planform approaches that of a swept wing of constant chord; there is, however, a rounding or filleting of the trailing edge at the root section.

The distribution of lift over the surface of wings having this form is illustrated in Fig. A,8d. It will be noted that a sharp drop in the lifting pressure occurs behind section AA and that no lift occurs behind the section BB at which the maximum span is reached. These discontinuities in the calculated results must of course be associated with the limiting case of zero width, and a more complete treatment, taking into account the three-dimensional nature of the flow, would show some redistribution of lift in these regions on wings of finite span.

As in the first example (Fig. A,8a), the exact shape of the tip and trailing edge behind the section of maximum span has no effect on the flow, nor on the total lift. The formulas for the lift and induced drag are

$$\frac{L}{\frac{1}{2}\rho_{\infty}U^{2}\alpha} = \frac{\pi}{2}\,(b^{2} - b_{1}^{2}) \tag{8-12}$$

$$\frac{D_{i}}{\frac{1}{2}\rho_{\infty}U^{2}\alpha^{2}} = \frac{\pi}{4}\,(b^{2} - b_{1}^{2}) - 2\frac{y_{1}}{b}\,[b^{2}E(k) - (b^{2} - b_{1}^{2})K(k)] \tag{8-13}$$

where b is the span of the wing (Fig. A,8c), b_{1} is the span of the wake at section BB, and $k = b_{1}/b$.

It is of interest to note that Eq. 8-12 is the result that would be obtained by applying the theory of oblique cylindrical flows (Art. 4) to an equivalent swept wing having a straight trailing edge (i.e. one which lacks the fillet area indicated on Fig. A,8c), at the same time taking into account the absence of lift in the tip region indicated by the slender wing theory. The area added by the curved trailing edge is evidently just sufficient to compensate for the interference between the two sides of the wing for every ratio of b_{1}/b.

Increasing the aspect ratio of a wing having a large angle of sweep does not necessarily increase the slope of the lift coefficient curve. For example, if one begins with a slender triangular wing, the aspect ratio may be made as large as desired by cutting out V-shaped areas, leaving a sweptback wing of constant chord. According to the foregoing formulas the lift curve slope would at first decrease, but would eventually, as the aspect ratio approached infinity, return to the value given by the original triangle. However, the ratio of lift to drag continually increases with aspect ratio.

Effect of compressibility. Application of the Prandtl transformation to the preceding formulas shows that in the limiting condition approached by the slender wing there is no effect of compressibility on the pressures. As explained in Art. 5, the incompressible flow corresponding to the compressible flow over a wing at Mach number M_{∞} is obtained by a calculation based on a fictitious wing whose lengthwise dimensions have been increased in the ratio $1/\sqrt{1 - M_{\infty}^{2}}$. The pressures at corresponding points on the actual wing are then obtained by increasing the values calculated for the fictitious wing in this same ratio. If the wing is sufficiently slender, however, the pressures will vary inversely as the length, and hence the

Prandtl transformation merely restores the pressures to their value at $M_\infty = 0$. As pointed out in [105], these statements apply to supersonic as well as subsonic speeds so long as the wing lies near the center of the Mach cone originating at its apex.

Application to the problem of flow in the vicinity of sonic speed. The velocity potential of the flow disturbance caused by a slender lifting surface of arbitrary form traveling at the velocity U may be represented by the function

$$\varphi = F(y + iz, x - Ut) \tag{8-14}$$

which satisfies the two-dimensional equation

$$\varphi_{yy} + \varphi_{zz} = 0 \tag{8-15}$$

in planes at right angles to the velocity of flight. In this form the pressure is given by

$$\Delta p = \rho_\infty \frac{\partial \varphi}{\partial t}$$

Eq. 8-14 represents a complete three-dimensional flow only in the sense of the approximate slender wing theory. When the velocity U is set equal to the velocity of sound a_∞, however, Eq. 8-14 becomes an exact solution of the sound wave equation in three dimensions (3-9). Thus the formulas developed for slender wings become applicable to wings of finite aspect ratio at speeds near the speed of sound [116,117]. This fact may be verified in another way by considering the flow to be steady, and applying the Prandtl-Glauert correction. The correction yields an equivalent wing whose aspect ratio is continually reduced as the Mach number increases toward 1.0.

At $U = a_\infty$, Eq. 8-14 represents a kind of sound wave, with lateral and vertical motions satisfying the boundary condition of the wing. The condition that the divergence of the component velocity field in the y, z planes be zero is analogous to the condition imposed by the one-dimensional flow equation (3-4), which prescribes no change in the cross-sectional area of the stream tubes at sonic velocity. Streamlines at sonic velocity are thus laterally incompressible.

An example of Eq. 8-14 is the family of solutions for wings of triangular planform, given by functions of the type

$$\frac{\varphi}{a_\infty} = (x - a_\infty t) F\left(\frac{y + iz}{x - a_\infty t}\right) \tag{8-16}$$

If the real solution is chosen and the substitution

$$\xi = \frac{y + iz}{x - a_\infty t}$$

is made, the expression

$$\frac{v - iw}{a_\infty} = F'(\xi) \tag{8-17}$$

for the lateral and vertical velocities is obtained. The function F' may be selected with the aid of examples given previously (Table A,2). The pressures are determined from the acoustic formula

$$\frac{\Delta p}{\frac{1}{2}\rho_\infty a_\infty^2} = \frac{2}{a_\infty^2} \frac{\partial \varphi}{\partial t} = 2[\xi F'(\xi) - F(\xi)] \tag{8-18}$$

The isobars are straight lines radiating from the point $x = a_\infty t$, as shown in Fig. A,8e.

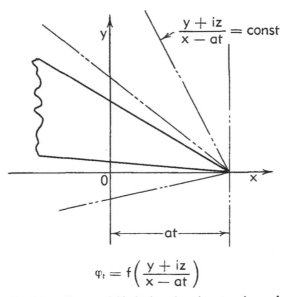

$$\varphi_t = f\left(\frac{y + iz}{x - at}\right)$$

Fig. A,8e.   Pressure field of triangular wing at sonic speed.

The application of the slender wing theory to flows in the vicinity of the velocity of sound has been shown [*116,117*] to be consistent with the assumptions of the linearized theory, provided the effects of thickness are excluded. Flows involving distributions of thickness generally include terms of the form

$$\varphi = \ln (y + iz) f(x - a_\infty t) \tag{8-19}$$

which lead to infinite pressure at infinity. The failure of the linearized theory in this case may be attributed to the presence of sources, which violates the condition of zero divergence. In the special case of the oblique cylindrical wing (Eq. 4-2) the total strength of the sources in each plane cross section is zero.

Nonlinear characteristics of low aspect ratio wings. Application of the linearized frictionless-flow theory to airfoils of low aspect ratio is limited more strictly to small angles of attack and small lift coefficients than is the case with airfoils of more usual proportions. As explained in Art. 4, the effect of yaw or sweep on a long narrow wing of cylindrical form is to reduce the lift coefficient for flow separation by the factor $\cos^2 \beta$. With relatively thin sections, such flow separation usually takes place at the leading edge as the result of the large pressure gradients associated with the flow around the nose of the section. With wings of low aspect ratio the flow in the vicinity of the leading edge remains locally similar to the flow around the edge of an oblique cylindrical wing, and hence flow separation can be expected at progressively smaller lift coefficients as the angle of sweep of the leading edge is increased. The linearized theory developed in the preceding sections is of course based on the assumption that the flow remains attached around the leading edge, and its validity is thus limited to a progressively smaller range of angles of attack as the leading edge is made more oblique. In the case of a sharp leading edge or a side edge parallel to the stream, separation can be expected at the smallest angles of attack.

Experimental observations of the flow after separation indicate the presence of a vortex surface which leaves the leading edge and rolls up in the form of a spiral sheet above and behind the edge. The radius of the spiral is small near the front of the wing and becomes larger at downstream sections. Partly as a result of the diffusion of vorticity, the spiral resembles a single vortex with a finite core. In this regime the lift continues to increase with angle of attack—frequently up to an angle of 45 degrees—but the variation is nonlinear and the lifting pressures are no longer divided equally between the upper and lower surfaces. Of the greatest practical consequence is the rapid increase of drag with angle of attack. After the flow becomes detached from the edge, the forward suction force no longer increases in proportion to the lift, with the result that the theoretical formulas for drag no longer apply and the resultant force on the wing falls back toward a direction at right angles to the chord plane. Prior to the occurrence of separation the drag is observed to follow roughly the theoretical minimum value

$$C_D = C_{D_0} + \frac{C_L^2}{\pi R}$$
(8-20)

but at higher angles of attack the value

$$C_D = C_{D_0} + C_L \tan \alpha$$
(8-21)

is approached.

Fig. A,8f, taken from data given in [*99*], illustrates the behavior observed experimentally in the case of low aspect ratio wings having different angles of obliquity of the leading edge. In each case the aspect ratio is $\frac{4}{3}$ and the airfoil section is the NACA 0012.

It will be noted that with wings of low aspect ratio the transition from one law of drag variation to the other is more gradual than in the case of the slender sweptback wing (Fig. A,4k).

A satisfactory quantitative treatment of the forces experienced by a wing in the nonlinear range is not yet available, though the behavior illustrated in Fig. A,4k and A,8f may be used as a guide. A treatment of the limiting case of a rectangular wing of small aspect ratio was given by Bollay [118]. Bollay assumes that a vortex sheet leaves the side edge of the wing, forming a region of constant negative pressure above the upper

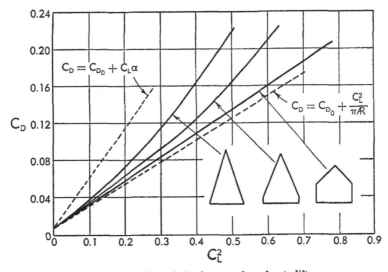

Fig. A,8f.   Effect of planform on drag due to lift.

surface. In the limiting case the lift curve is found to be parabolic, having zero slope at zero angle of attack.

The rectangular wing of zero aspect ratio might be treated by the theory of oblique cylindrical flows, the crosswise component of velocity being $U \sin \alpha$. The appropriate form of the two-dimensional crossflow in this case would be of the discontinuous type, such as the Helmholtz flow, satisfying the condition of tangential velocity at the side edges. Unfortunately there exists an infinite number of solutions of this type depending on the pressure assumed in the region above the wing, and there seems to be no physical criterion for selecting a certain value of this pressure. In Helmholtz' configuration the pressure is zero, but experimentally a considerable negative pressure is observed. A result in fair agreement with experiment in the case of a sufficiently long narrow plate can be obtained by calculating the force normal to the plate from an experimentally determined drag in a perpendicular air stream, corresponding to the component $U \sin \alpha$ [119,120]. The crosswise drag coefficient of such a plate has approximately the value 2.0, which leads to the expression for the normal force coefficient

$$C_n = 2 \sin^2 \alpha \qquad (8\text{-}22)$$

A,9. Thickness Distributions. In addition to the effects of camber and twist of the mean surface, consideration of the thickness distribution is required for the complete determination of the pressures and forces on the wing. In the linearized theory the thickness modifies the pressure distribution by the addition of equal pressures above and below the chord plane and hence does not affect the lift. Furthermore, in steady potential flow at subsonic speeds the thickness does not introduce drag. The additional flow velocities and pressure gradients arising from the thickness are decisive, however, in determining the Mach number at which the breakdown of smooth regular flow occurs and the subsequent rise in drag on transition to supersonic speeds.

AERODYNAMIC PROPERTIES OF THIN, FLAT ELLIPSOIDS. The potential flow over ellipsoids of various proportions can be completely determined by relatively simple formulas. Since a thin flat ellipsoid bears a close resemblance to a wing, a study of the flow and pressure distributions over such forms is of considerable interest in aerodynamics. With incompressible flow, the theoretical solutions are not limited by the assumptions of the thin airfoil theory and hence their study affords an understanding of certain errors involved in that theory. For application to high speeds the study of the incompressible case is made on the supposition that the aspect ratio and the velocity distribution are to be corrected by the Prandtl rule. In this application the approximations of the linearized theory must be borne in mind. Under most conditions the greatest departures from the linearized theory will occur near the rounded nose and tail of the elliptical sections and in these regions the effect of Mach number on the pressures will be less than indicated by the linearized theory.

For complete mathematical details of the potential flow calculations for ellipsoids, reference should be made to Lamb [*1*, pp. 139–156], or to [*121*, Vol. 1, pp. 293ff.] However, the results of the theory are of such simplicity that they can be completely summarized with the aid of a few elementary concepts.

As in the case of the lifting elliptic disk, the flow around nonlifting ellipsoids is studied with the aid of the ellipsoidal coordinates λ, μ, ν (Art. 7) and the associated Lamé functions. Since the vortex wake is absent, no simplification results from the use of the acceleration potential and hence the solutions are expressed in terms of the velocity potential

φ. Accordingly,

$$\varphi_n^m = E_n^m(\mu)E_n^m(\nu)F_n^m(\lambda) \tag{9-1}$$

where E_n^m and F_n^m are defined as in Art. 7.

The various solutions represented by Eq. 9-1 may be thought of as flows caused by various distributions of velocity around the surface of the base ellipsoid $\lambda = 0$. The solutions for uniform translation of the solid ellipsoid in the directions of x, y, and z are given by

$$\left.\begin{aligned}
\varphi_1^1 &= A_1^1 E_1^1(\mu)E_1^1(\nu)F_1^1(\lambda) = xf_1^1(\lambda) \\
\varphi_1^2 &= A_1^2 E_1^2(\mu)E_1^2(\nu)F_1^2(\lambda) = yf_1^2(\lambda) \\
\varphi_1^3 &= A_1^3 E_1^3(\mu)E_1^3(\nu)F_1^3(\lambda) = zf_1^3(\lambda)
\end{aligned}\right\} \tag{9-2}$$

On the surface of the base ellipsoid ($\lambda = 0$), $f(\lambda) = $ const, and it is seen that in each case the potential is proportional to the coordinate of the surface in the direction of motion. Hence the equipotential lines on the surface are ellipses formed by the intersections of a series of equidistant parallel planes with the surface $\lambda = 0$. For motion in any of the three principal directions, the planes of the equipotential lines are perpendicular to the direction of motion.

The form of the surface potential (9-1) leads to a simple rule for computing the velocity distribution over the surface of any ellipsoid fixed in an incompressible stream [9]. For this purpose it is necessary to know the magnitude and direction of the maximum velocity. The velocity at every point on the surface can then be obtained simply by taking the projection of the maximum velocity on the plane parallel to the surface at the point. For flow parallel to a principal axis, the maximum velocity is parallel to the direction of flow at infinity and appears all around the widest section of the ellipsoid. In oblique flow the maximum velocity will be inclined to the main flow direction; it is usually more convenient to treat oblique motion by resolving the velocity at infinity into components along the principal axes of the ellipsoid and superimposing the resulting velocity distributions.

The maximum velocity can be found directly from the coefficient of x, y, or z in the expression for φ, or in other words from the constants $f_n^m(0)$ in Eq. 9-2. In a stream of velocity U along any one of the three directions, the maximum velocity on the surface of the ellipsoid will be given by

$$[1 + f_n^m(0)]U$$

Hence

$$f_n^m(0) = \left(\frac{u}{U}\right)_{\text{max}}$$

The evaluation of the constants $f_n^m(\lambda)$ has been given in several text books [1; 121, pp. 224–293] and need not be reproduced here.

Fig. A,9a, plotted from data given in [122], shows values of $(u/U)_{max}$ for ellipsoids of various proportions. It will be noted that for every elliptic cylinder the maximum additional velocity is exactly equal to the thickness-chord ratio. Thus, for a thickness-chord ratio of 10 per cent and flow parallel to the chord plane, the velocity is $1.10U$, while for flow perpendicular to the chord plane the velocity around the nose and tail of the section is $11.0U$.

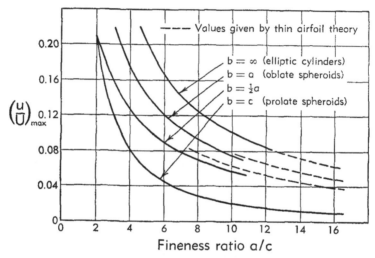

Fig. A,9a. Maximum velocity increments for ellipsoids; $\alpha = 0$, $M_\infty = 0$.

Of particular interest in the thin wing theory are the limiting values of $(u/U)_{max}$ for small thickness, i.e. $c/a \rightarrow 0$ and $c/b \rightarrow 0$. These limiting values are found to be

$$\left(\frac{u}{U}\right)_{max} = \frac{c}{a}\frac{ab(K-E)}{a^2-b^2} \quad \text{for } a > b \tag{9-3}$$

$$\left(\frac{u}{U}\right)_{max} = \frac{c}{b}\frac{a^2E-b^2K}{a^2-b^2} = \frac{c}{a}\frac{b^2E-a^2K}{b^2-a^2} \quad \text{for } b > a \tag{9-4}$$

$$\left(\frac{w}{U}\right)_{max} = \frac{a}{c}\frac{1}{E} \tag{9-5}$$

for flow in the direction of c normal to the chord plane. Here K and E are the complete elliptic integrals with modulus $k = \sqrt{1-(b/a)^2}$ or $k = \sqrt{1-(a/b)^2}$. Eq. 9-5 yields an infinite velocity around the edge of a perfectly thin disk ($c = 0$), but the value differs from that for two-dimensional motion by the factor $1/E$, which is the ratio of the span $2b$ ($b > a$) to its semiperimeter.

In flow parallel to the chord plane, as indicated in Fig. A,9a, the values of the maximum velocity given by the thin airfoil approximation are exact for the case of infinite aspect ratio ($b = \infty$), but are larger than the actual surface velocities for ellipsoids of finite proportions. The thin airfoil approximation cannot of course be used for a body of revolution since the velocities along the axis of singularities are infinitely greater than the surface velocities.

Fig. A,9b, taken from data given in [123], shows the effect of Mach number on $(u/U)_{max}$ for ellipsoids of various aspect ratios ($R = 4a/\pi b$) having thickness-chord ratios of 10 per cent. The calculations, which were based on the thin airfoil theory, show that the effect of Mach number is greatest in two-dimensional flow. Also indicated on the figure are values of the Mach number for which the linearized theory predicts the occurrence of a local supersonic region in the flow. The occurrence of such a region usually precedes the appearance of a shock, although no definite significance can be attached to such a calculation in three-dimensional flow. For the elliptic cylinder placed at an angle of yaw β, the designation of the abscissa may be replaced by $M_\infty \cos \beta$ and the quantities u_{max} and U will refer to the component velocities at right angles to the axis of the cylinder.

FLOW OVER WINGS OF CONICAL OR CYLINDRICAL FORM. The pressure field of thin nonlifting airfoils of rectilinear planform and simple thickness distribution can be obtained by direct integration of the effect of a distribution of sources over the chord plane. Calculations based on this method have been given in [124] and more recently in [125,126].

The method of [124] involves the use of the acceleration potential (Art. 7). In these terms a source represents a singularity in the pressure field or in what amounts to the same thing, the field of horizontal perturbation velocities $u(x, y, z)$. Thus the fundamental solution

$$\frac{u}{U} = -\frac{1}{r} = -\frac{1}{\sqrt{x^2 + y^2 + z^2}} \tag{9-6}$$

represents a point source of pressure rather than a source of volume.

To obtain the effect of a row of such sources—or a line source—along the x axis between the points a and b, it is necessary to integrate Eq. 9-6 as follows:

$$\frac{u}{U} = -\int_a^b \frac{d\xi}{\sqrt{(x - \xi)^2 + y^2 + z^2}}$$

$$= -\sinh^{-1}\frac{x - b}{\sqrt{y^2 + z^2}} + \sinh^{-1}\frac{x - a}{\sqrt{y^2 + z^2}} \tag{9-7}$$

The pressure field of the finite line source thus consists of the sum of two conical pressure fields radiating from the ends of the line source as in Fig. A,9c. Such a conical field with rays emanating from the origin can also be

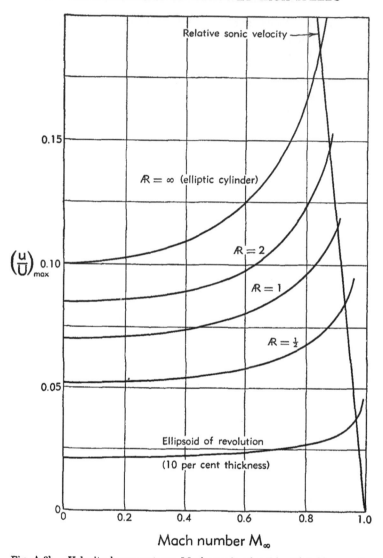

Fig. A,9b. Velocity increments vs. Mach number for thin ellipsoids; $\alpha = 0$.

represented in terms of the general solution of Laplace's equation of zero degree (Eq. 7-27); i.e.

$$\frac{u}{U} = \text{R.P.}\ln\frac{y+iz}{x+\sqrt{x^2+y^2+z^2}} = -\sinh^{-1}\frac{x}{\sqrt{y^2+z^2}} \quad (9\text{-}8)$$

If the velocity of flight is along the axis of the source, the flow will satisfy the boundary condition for a body of revolution. However, if the line source is turned out to a position oblique to the stream, the boundary shape will be distorted, and if the angle of obliquity is large enough to place the line source well outside the diameter of the original body the figure formed will be an oblique wedge. The nose angle of the wedge is formed where the streamlines of the main flow cross the line source.

The turning of the line source may be accomplished by the transformation

$$\left.\begin{aligned} x' &= x + my \\ y' &= y - mx \\ z' &= z\sqrt{1 + m^2} \end{aligned}\right\} \tag{9-9}$$

where m is the slope of the new axes relative to the old.

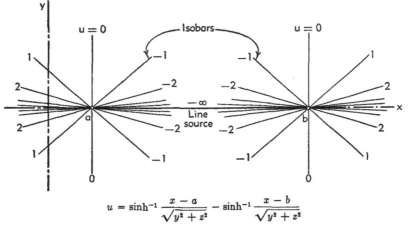

$$u = \sinh^{-1}\frac{x - a}{\sqrt{y^2 + z^2}} - \sinh^{-1}\frac{x - b}{\sqrt{y^2 + z^2}}$$

Fig. A,9c. Radial isobars from ends of line source

The vertical velocity w near $z = 0$, which determines the shape of the boundary, may be found by integrating u with respect to x and then differentiating the resulting velocity potential with respect to z:

$$w = \frac{\partial \varphi}{\partial z} = \frac{\partial}{\partial z}\int_{-\infty}^{x} u\, dx \tag{9-10}$$

Evaluation of this integral for the overlapping fields from the two ends of a line source gives

$$\frac{w}{U} = \pm 2\pi \frac{\sqrt{1 + m^2}}{|m|} \tag{9-11}$$

for the region of the x, y plane behind the line source. The plus and minus signs refer to the upper and lower surfaces of the chord plane, respectively. The figure formed by the streamlines crossing a line source is thus a wedge-shaped body having an oblique leading edge and extending in-definitely downstream.

By making use of the relation

$$\frac{\partial z}{\partial x} = \frac{w}{U}$$

together with Eq. 9-7 and 9-11, the pressure coefficient near the plane $z = 0$ may be expressed in terms of the slope as follows:

$$\frac{\Delta p}{\frac{1}{2}\rho_\infty U^2} = \frac{1}{\pi} \frac{|m|}{\sqrt{1 + m^2}} \frac{dz}{dx} \left(\sinh^{-1} \frac{x' - b'}{|y'|} - \sinh^{-1} \frac{x' - a'}{|y'|} \right) \quad (9\text{-}12)$$

Airfoils bounded by plane surfaces. It has been seen that the effect of a line source in the pressure field is to cause a deflection of the stream-lines crossing the source in such a way as to spread the streamlines apart. If the source is followed by a sink of equal strength, an equal and opposite deflection of the streamlines will occur as they cross over the sink. The figure formed by the streamlines near the plane $z = 0$ will thus be a plate of uniform thickness with a beveled leading edge, as shown in Fig. A,9d.

The pressure distribution over such a beveled edge may be obtained very simply by superimposing the pressures marked off on radial isobars originating from the four corners of the bevel. Fig. A,9d illustrates this process for a bevel having a square planform. Only isobars from one tip are shown because of the symmetry of the figure. With the corners at ± 1 and source and sink lines parallel to the y axis, we have

$$u = -\sinh^{-1} \frac{y + 1}{|x + 1|} + \sinh^{-1} \frac{y - 1}{|x + 1|} + \sinh^{-1} \frac{y + 1}{|x - 1|}$$
$$- \sinh^{-1} \frac{y - 1}{|x - 1|} \quad (9\text{-}13)$$

If the aspect ratio of the figure is increased, the isobars in intermediate regions approach parallel straight lines and the flow field thus approaches a cylindrical or two-dimensional form. Suppose that the tips are at $y = \pm \eta$, with η very much larger than 1. Then for the center of the wing, the terms $\sinh^{-1}(y \pm \eta)/(|x \pm 1|)$ in the expression for the pressure approach the limiting values

$$\sinh^{-1} \frac{y \pm \eta}{|x \pm 1|} \rightarrow \pm \ln 2 \left| \frac{y \pm \eta}{x \pm 1} \right|, \qquad y \ll \eta \quad (9\text{-}14)$$

with the sign of the logarithm determined by the sign of $y \pm \eta$, and the right-hand side of Eq. 9-13 is found to approach the Legendre function Q_0, i.e.

$$u \rightarrow -2 \ln \left| \frac{x - 1}{x + 1} \right| = 4Q_0(x) \quad (9\text{-}15)$$

This expression when combined with Eq. 9-11 agrees with the two-dimensional complex velocity function for the wedge

$$(u - iw) = 4Q_0(x) \mp 2\pi i P_0(x) \tag{9-16}$$

(see formula 19, Table A,2).

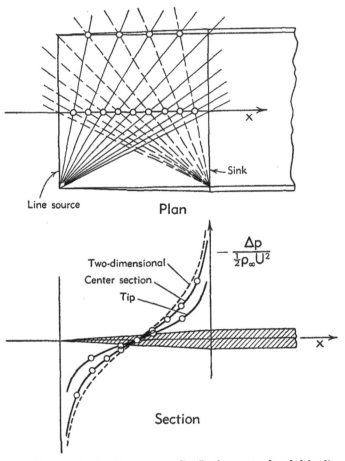

Fig. A,9d. Construction for the pressure distribution over a beveled leading edge.

Eq. 9-15 applies over the middle portion of a long narrow wedge. At the extreme tips it may be shown that the pressure falls to one half this value. Although the thin airfoil theory indicates that the thickness of the figure ends abruptly at the ends of the source distribution, a more exact consideration would be expected to show some rounding of the tip.

Airfoils of biconvex section. The use of a finite number of source-sink lines results in airfoil sections composed of straight segments. Cylindrical surfaces having continuous curvature require a continuous distribution of sources and sinks aligned with the generators of the surface. The simplest of these is the biconvex profile in which the upper and lower surfaces are parabolic arcs of constant curvature. Such a profile requires line sources of finite strength to form the desired angle of intersection of the arcs at the leading and trailing edges, together with a uniform distribution of sinks between the two sources.

The pressure field for a uniform sheet of line sources is obtained by integrating in the x direction the pressure field of a single line source. The result is

$$\int \sinh^{-1} \frac{x'}{|y'|}\, dx = \frac{\sqrt{1+m^2}}{m}\, y \sinh^{-1}\frac{x}{|y|} - \frac{1}{m} y' \sinh^{-1}\frac{x'}{|y'|} \qquad (9\text{-}17)$$

For a bilaterally symmetrical arrangement,

$$\int \left(\sinh^{-1} \frac{x'}{|y'|} + \sinh^{-1} \frac{\bar{x}'}{|\bar{y}'|} \right) dx = \frac{1}{m} \left(\bar{y}' \sinh^{-1}\frac{\bar{x}'}{|\bar{y}'|} \right.$$
$$\left. - y' \sinh^{-1}\frac{x'}{|y'|} \right) \qquad (9\text{-}18)$$

where $\bar{x}'$ and $\bar{y}'$ have been used to denote the "conjugate" coordinates $x - my$ and $y + mx$, respectively.

To obtain a complete sweptback wing it is necessary to superimpose a number of component pressure fields, as explained in [126] and [127]. For a wing of infinite span with leading and trailing edges at $y' = +m$ and $-m$, respectively, on one side and $\bar{y}' = +m$ and $-m$ on the other side, the result is

$$\frac{\Delta p}{\frac{1}{2}\rho_\infty U^2} = \frac{2}{\pi} \left(\frac{t}{c}\right)_{\max} \frac{m}{\sqrt{1+m^2}} \left[\frac{y'}{m} \left(\sinh^{-1}\frac{x'+1}{|y'-m|} \right. \right.$$
$$\left. - \sinh^{-1}\frac{x'-1}{|y'+m|} \right) + \frac{\bar{y}'}{m} \left(\sinh^{-1}\frac{\bar{x}'-1}{|\bar{y}'-m|} \right.$$
$$\left. \left. - \sinh^{-1}\frac{\bar{x}'+1}{|\bar{y}'+m|} \right) + 2Q_1\left(\frac{y'}{m}\right) + 2Q_1\left(\frac{\bar{y}'}{m}\right) \right] \qquad (9\text{-}19)$$

where $(t/c)_{\max}$ is the thickness-chord ratio of the biconvex profile. The terms in Q_1 represent the pressure distribution on the airfoil in two-dimensional flow (see formula 20, Table A,2). The appearance of these terms is the result of the assumption that the tips are removed to a great distance.

Fig. A,9e shows the pressure distributions at various stations along the span for a biconvex wing having 60° of sweepback. The curves assume the two-dimensional form at a relatively short distance ($y > c/2$) from the root section; similar behavior is to be expected near the tips. Hence the assumption of infinite aspect ratio should apply very nearly at any section situated more than one-half chord length from either root or tip.

At the root section ($y = 0$), Eq. 9-19 reduces to

$$\frac{\Delta p}{\frac{1}{2}\rho_\infty U^2} = \frac{2}{\pi}\left(\frac{t}{c}\right)_{\text{max}} \frac{m}{\sqrt{1+m^2}}\left[4Q_1(x) - 4P_1(x)\sinh^{-1}\frac{1}{m}\right] \quad (9\text{-}20)$$

Since $dz/dx = -2(t/c)_{\text{max}}P_1(x)$, the second term in the bracket denotes a component of the pressure which is proportional to the local slope of the

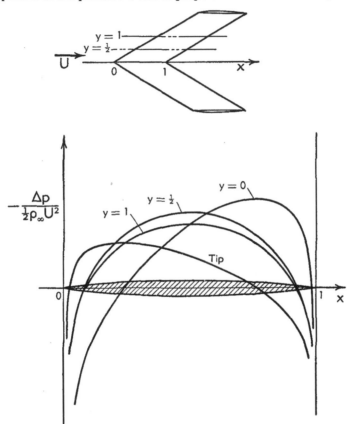

Fig. A,9e. Pressure distribution at various spanwise stations on a sweptback wing; $\Lambda = 60°$; $M_\infty = 0$.

airfoil surface. Such a component appears for a section of any shape and is a characteristic of the supersonic flow over a section. At subsonic speeds its magnitude increases progressively with the Mach number.

Fig. A,9f shows the effect of Mach number on the pressures over the root section as obtained by applying the Prandtl-Glauert transformation to the foregoing formula. It will be noted that an increase of the Mach number causes a progressive transition toward the supersonic, or Ackeret, type of pressure distribution ($M_\infty = 1.05$) to be discussed in a later section. Sections farther out along the span retain the subsonic type of pressure distribution. In the case of the straight wing, a more rapid change in the pressure distribution occurs on transition to supersonic speeds.

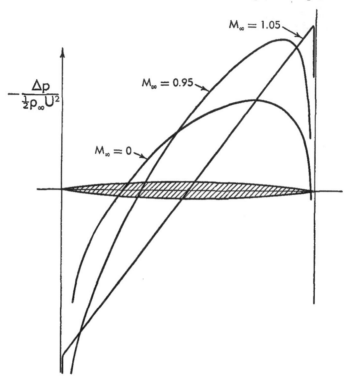

Fig. A,9f.   Effect of Mach number on the pressure distribution over the root section of a sweptback wing; $\Lambda = 60°$.

CHAPTER 3. AERODYNAMICS OF THIN WINGS AT SUPERSONIC SPEEDS

A,10.   Fundamental Considerations.

Geometric character of supersonic disturbances.   As the speed is increased in the subsonic range the disturbance field of the moving wing becomes progressively modified, the disturbance extending to greater distances both laterally and vertically. For steady motion this distortion is shown by the Prandtl-Glauert transformation, which prescribes a change in the scale of the lateral and vertical dimensions relative to the lengthwise dimension. Except for this continuous change in proportions, however, there remains an essential similarity to incompressible flow. The features of the incompressible flow, such as the increase of speed at the sides of a body, the recovery of pressure at the rear, and the fore-and-aft balance of the pressure forces (D'Alembert's paradox) remain. Hence the intuitive concepts acquired in the study of incompressible flow suffice almost without modification for the whole range of steady subsonic potential flow.

On transition to supersonic speeds a real change in the geometric character of the disturbance field takes place. The fore-and-aft symmetry of the subsonic flow disappears and the disturbance is limited to the region behind the wing bounded by Mach waves or cones. Because of the asymmetry of the flow, all outgoing disturbances result in a drag, so that the thickness of the body or wing creates a drag in addition to the drag associated with the lift. At higher supersonic speeds the lifting efficiency of the wing begins to be diminished by the restriction on the volume of air involved in the production of the lift. As the speed is increased in the supersonic range the angle of the Mach waves becomes more acute and the volume of air coming under the influence of the wing becomes progressively smaller until, finally, it includes only that mass of air actually encountered by the frontal projection of the wing area.

The additional drag associated with the supersonic speed has been termed the "wave drag" since it appears in the energy required to extend the wave system emanating from the body. From the standpoint of an observer at rest, the streamlines of the disturbance field originate at the fronts of these waves. The streamlines always intersect the wave fronts at right angles, with the result that the motion at the wave front is parallel to the direction of propagation, in accordance with the simplified concept of a sound wave as a "longitudinal" disturbance. In the vicinity of the wave fronts, discontinuous changes in pressure and velocity may occur, but in three-dimensional supersonic flows there also appear large regions in which the motion is continuous, and the streamlines are indistinguishable from the patterns of an incompressible flow.

According to the thin airfoil theory the wave drag is zero at all subsonic speeds and rises suddenly at $M_\infty = 1.0$. As shown by Busemann [*128*] this discontinuous behavior results from the idealization of steady flow. In nonstationary or curvilinear motion, such as the motion of a propeller blade, the wave drag begins to appear at subsonic speeds.

From the mathematical standpoint, the change in flow geometry on going to supersonic speed appears in the change of the disturbance equation from the elliptic to the hyperbolic type. The steady flow equation

$$(1 - M_\infty^2)\varphi_{xx} + \varphi_{yy} + \varphi_{zz} = 0 \qquad (10\text{-}1)$$

becomes on transition to supersonic speeds

$$(M_\infty^2 - 1)\varphi_{xx} - \varphi_{yy} - \varphi_{zz} = 0 \qquad (10\text{-}2)$$

The disturbance fields admitted as solutions of the latter partial differential equation show a remarkable variety of pattern. In three dimensions these patterns range from the straight and discontinuous wave fronts associated with the hyperbolic equation, to the intricate curvilinear geometry associated with solutions of Laplace's equation. This variety can be foreseen by examining arrangements of the three terms of the differential equation in pairs.

First let us normalize the equation by the transformation

$$x_1 = \frac{x}{\sqrt{M_\infty^2 - 1}}; \quad y_1 = y; \quad z_1 = z \qquad (10\text{-}3)$$

so that

$$\varphi_{x_1 x_1} - \varphi_{y_1 y_1} - \varphi_{z_1 z_1} = 0 \qquad (10\text{-}4)$$

Then, dropping the subscripts, we have for one pair of terms

$$\varphi_{xx} - \varphi_{zz} = 0 \qquad (10\text{-}5)$$

with solutions of the form

$$\varphi = F(x \pm z) \qquad (10\text{-}6)$$

corresponding to the flows considered in the Ackeret theory (see Art. 11). Such expressions can represent the wave disturbances in the vicinity of the sections of a long straight wing.

On the other hand, considering the last two terms of the equation, we have

$$\varphi_{yy} + \varphi_{zz} = 0 \qquad (10\text{-}7)$$

and the solutions have the form

$$\varphi = F(y \pm iz) \qquad (10\text{-}8)$$

i.e. the form of an incompressible potential flow in two dimensions. This is the form approached by the supersonic disturbance field in the vicinity of the vortex wake at a great distance behind the airfoil. In the case of a large angle of sweep, Eq. 10-7 may also apply in the vicinity of the wing itself.

Further illustration of the basic difference in geometry is shown by the fundamental solutions of the two flow equations, viz.

$$\varphi = \frac{1}{\sqrt{x^2 + (1 - M_\infty^2)(y^2 + z^2)}} \quad \text{(subsonic)} \qquad (10\text{-}9)$$

and

$$\varphi = \frac{1}{\sqrt{x^2 - (M_\infty^2 - 1)(y^2 + z^2)}} \quad \text{(supersonic)} \qquad (10\text{-}10)$$

These solutions, which represent the velocity potential arising from a point source, are illustrated in Fig. A,10a. In the subsonic case the equipotential surfaces are ellipsoids which can be reduced to spheres by the

Prandtl-Glauert transformation. In the supersonic case the surfaces are hyperboloids of two sheets bounded by the Mach cone, which extends both ahead of and behind the point source. Outside this cone the potential is imaginary. It is evident that the differential equation makes no distinction between forward and rearward zones of influence of the source, and the disturbance in either branch of the Mach cone complies with the differential equation. The upstream branch of the Mach cone is ordinarily to be discarded for physical reasons. Discarding the forward branch of the Mach cone, however, requires that the strength of the disturbance be doubled to provide the same flux from the source. The velocity distribution at the center of the Mach cone, along the positive branch of the x axis, is exactly twice that for a subsonic source.

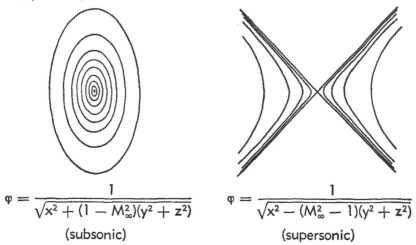

$$\varphi = \frac{1}{\sqrt{x^2 + (1 - M_\infty^2)(y^2 + z^2)}}$$

(subsonic)

$$\varphi = \frac{1}{\sqrt{x^2 - (M_\infty^2 - 1)(y^2 + z^2)}}$$

(supersonic)

Fig. A,10a. Equipotential surfaces for a subsonic and a supersonic source.

A physical explanation of this difference in the action of subsonic and supersonic sources may be adapted from the explanation given by Heaviside [*129*] for the case of a point charge moving at a speed less than or greater than the speed of light. A diagram similar to Heaviside's is given in Fig. A,10b. The field of the steadily moving source may be obtained by the superposition of the nonstationary fields of a series of pulses which go on and off in such a sequence as to represent the motion of the source along the line. Each pulse sends out a spherical wave which expands with the radial velocity a. In the subsonic case the expanding spheres fill the whole space without overlapping so that an observer at any point sees (or hears) the pulse traveling along in a retarded position. In the supersonic case, however, each point within the Mach cone is characterized by two intersecting spheres, one having its center along the line of motion ahead of the observer (P) and the other behind. Thus in the supersonic case the observer sees two sources whose apparent positions recede from him in opposite directions.

Wing flows related by the Prandtl-Glauert transformation. Steady flows at different Mach numbers in the supersonic range may be related geometrically by a transformation of coordinates similar to the Prandtl-Glauert transformation, but with the factor $1/\sqrt{1 - M_\infty^2}$ replaced by

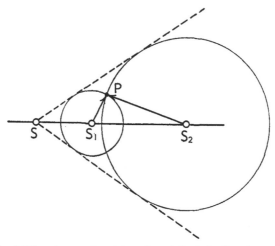

Fig. A,10b. Apparent positions S_1 and S_2 of the disturbance S traveling at supersonic speed. Observer is at P.

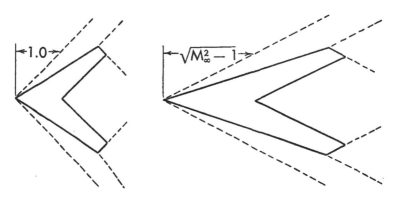

Fig. A,10c. Wings having flows related by the Prandtl-Glauert transformation.

$1/\sqrt{M_\infty^2 - 1}$. This is the transformation already introduced (Eq. 10-3). As illustrated in Fig. A,10c, it shortens or lengthens the potential field in accordance with the change in inclination of the Mach lines. In general, increases of the Mach number in the supersonic range decrease the ratio of the longitudinal to the vertical and lateral perturbation velocities by $1/\sqrt{M_\infty^2 - 1}$.

At the wave front the resultant disturbance velocity becomes normal to the wave in all cases, so that at such points

$$\frac{u}{\sqrt{v^2 + w^2}} = \frac{1}{\sqrt{M_\infty^2 - 1}} \tag{10-11}$$

At other points behind the wave front and within the zone of influence of the wing, the disturbance field presents a more-or-less complex three-dimensional pattern, tending generally toward the pattern of an incompressible flow near the center of the Mach cone as the zone of disturbance widens.

Subsonic and supersonic edges. Wherever the outline of the wing forms a continuous curve in plan view it is possible to mark off a sufficiently narrow strip around the edge such that the flow in the vicinity of this strip is locally cylindrical or two-dimensional in form. Hence the nature of the flow in the vicinity of the wing edges may be deduced from the theory of oblique cylindrical flows previously discussed (Art. 4). On this basis it is seen that an important distinction arises between cases in which the component velocity normal to the edge is subsonic and those in which this velocity is supersonic. In the former case the rules of subsonic flows will apply and hence, for brevity, it is said that the edge is subsonic. Similarly, whenever the normal component of velocity is supersonic, the edge is said to be supersonic.

Thus, in the case of a lifting wing having a rounded subsonic leading edge, there will arise a flow around the edge and a suction force will be developed, as in the Kutta-Joukowski flow (Art. 2). The effect of a symmetrical distribution of thickness, on the other hand, is to develop a positive pressure in this region, the maximum value of which will be equal to the stagnation pressure corresponding to the transverse velocity component at the edge.

If the leading edge is supersonic the flow will be locally similar to the Ackeret flow, provided the edge is sharp. In this case the flow is characterized by a curved shock wave attached to the edge. Here there will be no flow around the edge and the surface of the airfoil divides the flow into independent upper and lower regions. As in the Ackeret theory, it is no longer necessary to distinguish between pressures arising from the lift and those arising from the thickness, the pressures on each surface being determined by the shape of that surface alone. If the supersonic edge is blunt or rounded, the shock wave will be detached and the pressures cannot be calculated by the Ackeret theory. A satisfactory extension of the thin airfoil theory to deal with examples of this kind has not yet been devised.

In the case of a supersonic trailing edge, the independent pressures

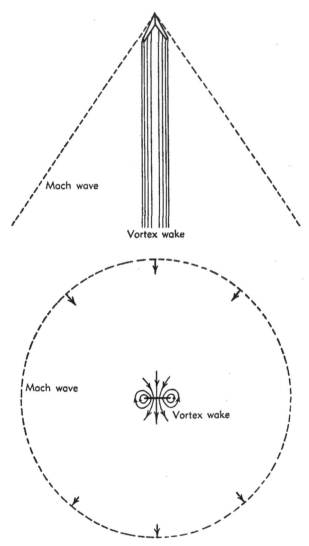

Fig. A,10d.   Disturbance produced by wing at supersonic speeds.

on the upper and lower surfaces will be maintained up to a point so near
the trailing edge that equalization of the pressures takes place through
the boundary layer. In theory it is usually considered sufficient to satisfy
the condition that the lift fall to zero behind the wing by discontinuous
pressure changes across oblique shock waves attached to the edge. In
practice, because of the existence of the viscous wake, the attachment of
the waves to the trailing edge is never complete and the result is that the

pressure changes in this region occur somewhat more gradually than the theory would indicate.

Wave drag; vortex drag. At supersonic speeds, drag arises both from the thickness of the wing and from the lift. The drag arising from the lift may be identified partly as vortex drag and partly as wave drag, though both effects appear in the rearward inclination of the lift vector. In the theory of wings at subsonic speeds it is shown that the production of lift by a wing of finite span is accompanied by a drag force that depends on the magnitude of the lift and its distribution over the span. In subsonic theory, this drag is termed the "induced drag" and may be calculated from consideration of the vorticity in the wake. It is evident that the total drag coefficient of a wing in supersonic flow cannot fall below the value of the induced drag coefficient corresponding to the same spanwise distribution of lift at subsonic speeds, since the vortex wake is similar and the motion induced by this wake at a great distance is similar to that in subsonic flow. At a great distance behind the wing, as pointed out by Hayes [44], the wave disturbance becomes concentrated in the vicinity of the Mach cone and does not modify the purely subsonic field of disturbance in the vicinity of the vortex wake, which occupies a relatively small region near the center of the cone (see Fig. A,10d). Hence the wave drag and the vortex drag may be separated if the calculation is made on the basis of the momentum of the flow at a great distance behind the wing.

For many purposes it is convenient to determine the drag from the pressures acting directly on the wing. In this way it is readily shown that the drag arising from the thickness is directly proportional to the square of the thickness-chord ratio, while the drag arising from the lift is proportional to the square of the lift coefficient. Changing the angle of attack or camber does not change the drag arising from the thickness, however, and hence the components of drag due to lift and thickness are additive in the approximation of the linear theory, just as in subsonic flow. These other components of the drag will be treated separately in the articles to follow.

A,11. Two-Dimensional Flow.

Ackeret theory. The first theoretical treatment of airfoils traveling at supersonic speed was given by Ackeret [10]. Ackeret considered the two-dimensional flow produced by a cylindrical wing traveling at right angles to its long axis. The sections were assumed to be thin, with sharp leading and trailing edges, so that the linearized equation of motion would apply. The disturbance field produced by the wing in this case turns out to be kinematically extremely simple. As illustrated in Fig. A,11a it consists merely of two plane waves of sound, traveling in the direction normal to the wave fronts and intersecting the x, y plane at the Mach angle μ

$= \sin^{-1}(1/M_\infty)$. Two waves are needed, one for the upper surface of the airfoil and one for the lower. The velocity with which the intercept travels along the x axis is greater than the actual velocity of the wave and is equal to the velocity of flight U. The equation of motion (10-6), which is the equation for plane sound waves, admits an arbitrary variation of velocity within the wave, but requires that the resultant disturbance velocity be at right angles to the wave front. Hence the horizontal and

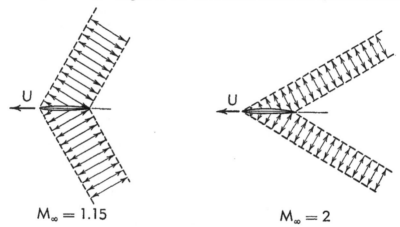

$M_\infty = 1.15$ $\qquad\qquad\qquad$ $M_\infty = 2$

Fig. A,11a. Ackeret flow, showing disturbance velocities produced by wing.

vertical components u and w remain in a fixed ratio, which depends on the Mach angle. A simple resolution of velocities shows that

$$u = w \cot \mu = \frac{w}{\sqrt{M_\infty^2 - 1}} \tag{11-1}$$

The variation of velocity within the waves is determined by the shape of the airfoil surface through the relation

$$\frac{w}{U} = \frac{dz}{dx} \tag{11-2}$$

as in the thin airfoil theory for subsonic speeds. The pressure disturbance is again given to the first order by the relation

$$\frac{\Delta p}{\frac{1}{2}\rho_\infty U^2} = -\frac{2u}{U} \tag{11-3}$$

The latter formula agrees with the pressure relation used in acoustical theory. Combination of the three equations yields an expression for the pressure coefficient in terms of the slope of the airfoil surface:

$$\frac{\Delta p}{\frac{1}{2}\rho_\infty U^2} = \frac{2(dz/dx)}{\sqrt{M_\infty^2 - 1}} \tag{11-4}$$

The pressure is constant at all points of any plane parallel to the wave front. The pressure at any point on the airfoil surface depends solely on the slope of the surface at that point and there is no influence of one portion of the airfoil on another.

With the aid of Eq. 11-4 it is easily shown that the lift and pressure drag per unit span of an airfoil are given by

$$L = -\tfrac{1}{2}\rho_\infty U^2 \frac{2}{\sqrt{M_\infty^2 - 1}} \int \frac{dz}{dx}\, dx \qquad (11\text{-}5)$$

$$D = \tfrac{1}{2}\rho_\infty U^2 \frac{2}{\sqrt{M_\infty^2 - 1}} \int \left(\frac{dz}{dx}\right)^2 dx \qquad (11\text{-}6)$$

where the integrations extend over both upper and lower surfaces of the airfoil. The highest ratio of lift to drag is achieved when dz/dx is a constant, i.e. when the airfoil is a perfectly thin flat plate. In this case the formulas, in coefficient form, are

$$C_L = \frac{4\alpha}{\sqrt{M_\infty^2 - 1}} \qquad (11\text{-}7)$$

$$C_D = \frac{4\alpha^2}{\sqrt{M_\infty^2 - 1}} \qquad (11\text{-}8)$$

As the latter formula shows, the lift is inclined backward at an angle equal to the angle of attack, and no suction force appears at the leading edge.

Higher approximations in the calculation of two-dimensional flows. At finite values of the thickness or angle of attack, departures from the linearized theory arise. Such departures make their appearance partly as an alteration in the geometry of the disturbance pattern, and partly through the nonlinear relation between the local velocity and the pressure. The former effect may be visualized as the result of the changes in the local inclination of the Mach lines in the field of varying local velocities. The nonlinear relation between the velocity and the pressure arises in part from the variation of density. The effect of a given pressure gradient in decelerating the stream is diminished by the accompanying increase of density along the streamlines. Similarly, the acceleration, in flowing toward a region of reduced pressure, is magnified by the reduction of density.

In two-dimensional flow, calculations can be made which show the nonlinear behavior of a perfect gas with a high degree of accuracy. The method of calculation, known as the "shock-expansion method" [130] makes use of the relations governing finite shock and expansion waves given originally by Hugoniot [131] and Meyer [132]. For a full explanation of this method, reference should be made to Ferri's book [133].

For airfoils of reasonable thickness the shock-expansion method shows an essentially single-valued, though nonlinear relation between the slope of the surface and the pressure. This relation is illustrated in Fig. A,11b for two values of the stream Mach number. It is seen that positive inclinations of the surface have a larger effect than predicted by linearized theory while negative inclinations have a smaller effect. In computations of the over-all lift or drag these differences tend to compensate each other.

Fig. A,11c shows pressure distributions calculated by the shock-expansion method for biconvex circular arc profiles of varying thickness ratio at a Mach number of 2.0. The comparison with linearized theory

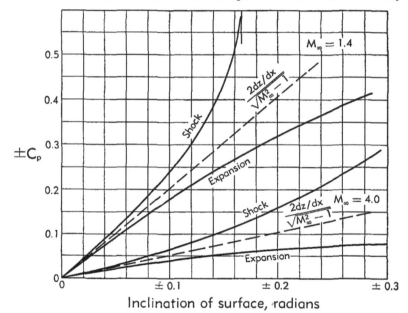

Fig. A,11b.   Pressure coefficient $C_p = (p - p_\infty)/\frac{1}{2}\rho_\infty U^2$ vs. surface inclination in two-dimensional flow.

shows good agreement at $t/c = 5$ per cent, but increasing departures for 10 per cent and 15 per cent thickness.

In a series of experiments made in the supersonic wind tunnel at Guidonia, Ferri [134] measured the pressure distributions for a variety of airfoil sections. The observed pressure distributions were in good agreement with those calculated by the shock-expansion method over the forward parts of the sections but the pressures near the trailing edge were considerably modified by flow separation. A typical result of Ferri's experiments is shown in Fig. A,11d. Here the airfoil was of biconvex section. Evidently in the case of 10 per cent thickness the pressure rise at the trailing edge of the airfoil was sufficiently great to cause a reversal of the boundary layer flow at the trailing edge. In the experiments quoted, the Reynolds number was rather lower than full size values, and the

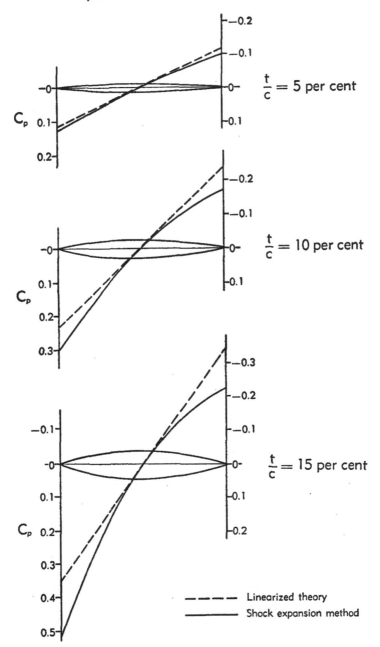

Fig. A,11c.   Comparison of pressures given by linearized theory and shock-expansion method; two-dimensional flow, $M_\infty = 2.0$.

boundary layer over the airfoil was laminar. At higher Reynolds numbers the boundary layer is probably turbulent and flow separation is less likely to occur. In spite of the difference in pressure distribution, the relation between airfoil thickness and flow separation appears to be much the same at supersonic speeds as at low subsonic speeds.

Optimum profiles. Because of the magnitude of the theoretical pressure drag in two-dimensional supersonic flow, the occurrence of separation actually may have the effect of diminishing the drag in certain cases. In such cases it is desirable to thicken the airfoil at the trailing edge and to

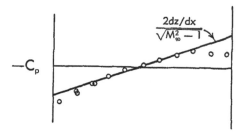

Pressure distribution

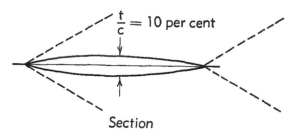

Section

Fig. A,11d. Comparison of experimental and theoretical pressure distributions; $M_\infty = 2.13$, $Re = 640,000$. Data from [*134*].

diminish the slope over the forward part [*135,136,137,138*]. The wave drag is thus lessened while the drag appearing in the wake is increased.

The optimum shape of the profile, taking into account the trailing edge thickness, is not readily determined. However, if the thickness-chord ratio and one other (e.g. structural) condition are specified, a shape for minimum pressure drag can be calculated [*137,138*], provided that a value for the base pressure can be estimated. Using experimentally determined values for the base pressure, Chapman [*137,138*] found that the greater the thickness ratio, or the higher the Mach number, the greater the relative bluntness required at the trailing edge, although the exact thickness is more critical in the case of thin airfoils and moderate Mach numbers. It is interesting to note that essentially the same results were obtained both by linear and shock-expansion methods.

Oblique cylindrical flows at supersonic speed. In view of the continuous decrease of efficiency of the Ackeret type of flow with increased Mach number, Busemann [*32*] suggested that the wings be swept back so as to reduce the component Mach number of the wing sections to a lower value. Busemann showed that a sufficiently long wing, swept so that its crosswise velocity was only slightly supersonic, was aerodynamically more efficient than a straight wing.

In Busemann's analysis, the flow over the oblique wing panels was assumed to be two-dimensional or cylindrical in form, so that the solution could be obtained by superimposing an ineffective axial component of velocity on a flow of the Ackeret type. The field of disturbance in this case may be visualized as a formation of plane waves inclined both vertically and laterally in such a way that the lines of intersection with the x, y plane travel at the crosswise velocity $U \cos \Lambda$, while the point of intersection with the x axis travels at the flight velocity U. The relations governing the lift and drag remain the same as in straight flow (Eq. 11-5 and 11-6), provided the reduced values of Mach number and velocity are employed. It is important to note, however, that the pressure drag determined in this way is directed at right angles to the leading edge and hence its effectiveness in opposing motion in the direction of flight is reduced. Since the effectiveness of the lift force is not similarly reduced, it is evident that this inclination of the pressure drag is advantageous, although the advantage is partly offset by the fact that the friction drag force is still essentially governed by the resultant flight velocity U.

Later calculations [*33*] have indicated that a still greater improvement in aerodynamic efficiency may be obtained by increasing the angle of sweep, so that the wing lies inside the Mach cone and the crosswise component of velocity becomes subsonic. In the case of the infinite wing, the Ackeret type of flow is then replaced by a flow of the Kutta-Joukowski type and the wave drag disappears. The relations determining the flow and the forces on the two-dimensional airfoil with a subsonic normal velocity are given in Chap. 2. For a wing of finite span, however, the wave drag does not disappear entirely when the wing is swept behind the Mach cone. Determination of the wave drag requires consideration of the three-dimensional flow field, as described in the following articles.

A,12. The Drag of Lifting Surfaces in Three-Dimensional Flow.

Calculation from lift distribution. At supersonic speeds, the form of the vortex wake following a given distribution of lift, as well as the induced field of motion in the vicinity of the wake far behind the wing, is the same as at subsonic speeds. Hence, the vortex drag (Art. 10) of a wing traveling at supersonic speed can be identified with the induced drag which would result from a similar distribution of lift at subsonic speeds. Following Munk's "stagger theorem," this component of the drag will depend only on the spanwise distribution of lift and will be independent of the distribution of lift along the chord of the wing. However, the wave drag—i.e. the component of drag associated with the energy in the system of waves continuously being extended by the airfoil cannot be related simply to spanwise or lengthwise loading at the wing, but depends in a more complex way on the actual distribution of lift over the surface. The total drag may be found by integrating the product of the lift and the downward inclination of the streamlines over the airfoil surface; this method is usually most convenient when the geometry of the airfoil is simple (e.g. in the case of a flat wing).

In computing the total drag by integrating around the surface, care must be taken to include the alleviating effect of the suction force along any subsonic leading edges (see Art. 10). The general procedure described in Art. 2, applied to the three-dimensional problem, will reduce in the limit of zero thickness to the consideration, for each infinitesimal segment of the leading edge, of local conditions in the plane normal to the edge.

Thus, if the incremental velocity over the surface, normal to the leading edge, includes a singularity in the form $C\Delta^{-\frac{1}{2}}$, where Δ is the length of the normal to the leading edge, the leading edge suction force will be given by Eq. 2-14.

Eq. 2-14 is derived for incompressible flow. For supersonic speeds it will be necessary to take into account the effect of compressibility by means of the Prandtl-Glauert transformation, and in addition, for practical convenience, it will be desirable to rewrite the formula in terms of the usual flight-oriented coordinates. The quantity of interest is the component of the suction force in the flight direction, or the thrust. If M_∞ is the free stream Mach number and ρ_∞ the density, and m is the ratio of the local inclination of the leading edge to the tangent of the Mach angle ($\sqrt{M_\infty^2 - 1}$/tangent of the angle of sweep), the thrust per unit streamwise length of leading edge may be written

$$\frac{dT}{dx} = \pi\rho_\infty \sqrt{1 - m^2}\, C_x^2 \qquad (12\text{-}1)$$

where C_x is the value at the leading edge $x = x_{1.e.}$ of $u \sqrt{x - x_{1.e.}}$, x and u being measured in the streamwise direction.

In the following subarticle an alternate method of calculating the drag arising from a given distribution of lift at supersonic speeds will be described. This method has the advantage of showing explicitly the type of lift distribution that leads to small values and, in certain cases, to the minimum value of the drag.

At subsonic speeds the drag is a minimum, for a given lift and a given span, when the induced downwash (in the special sense of the lifting line theory) has the same value at all points of the span. In Art. 6 it was shown that the minimum drag consistent with a given planform and a given total lift occurs when the superposition of the forward and reversed disturbance fields results in a constant value of the downwash at all points of the planform. In the more general theory, which includes both subsonic and supersonic speeds, the "combined downwash" thus takes the place of the induced downwash.

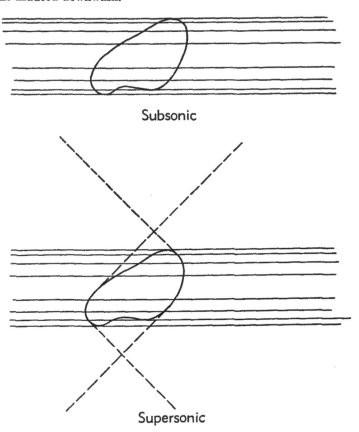

Fig. A,12a.   Lifting surfaces with superimposed disturbance fields.

The superposition of the forward and reversed disturbance fields, corresponding to a given distribution of lift, results in a purely two-dimensional field of motion if the flight speed is subsonic. As illustrated in Fig. A,12a, the combined disturbance field in this case is obtained simply by extending the trailing vortex wake straight across the wing into the region ahead. Since the problem is two-dimensional, the determination of the optimum distribution of lift is relatively simple.

At supersonic speeds the combined disturbance field is again characterized by a vortex ribbon, identical in form with the trailing vortex wake, and extending across the wing both ahead of and behind the wing. Here, however, the field of motion is three-dimensional in form and is bounded by two overlapping zones of influence, as illustrated in the lower part of the figure. The determination of the ideal distributions, which yield a constant value of the downwash $\bar{w}$ at each point of the planform, thus leads to a considerably more difficult calculation if the speed is supersonic.

Plane-waves method. The method to be presented for the calculation of the drag makes use of forms similar to Whittaker's solution of Laplace's equation [52] and is related to a more general class of integral operators studied by Bergman [139]. Starting with the potential function

$$\varphi = F(X) \tag{12-2}$$

where

$$X = \alpha x - \beta y - \gamma z \tag{12-3}$$

we find that φ is a solution of the normalized potential equation (10-4) if F is a differentiable function and if α, β, and γ are parameters determined so that

$$\alpha^2 - \beta^2 - \gamma^2 = 0 \tag{12-4}$$

Setting

$$\left. \begin{aligned} \alpha &= 1 \\ \beta &= \cos \theta = \frac{1}{2}\left(\frac{1}{\lambda} + \lambda\right) \\ \gamma &= \sin \theta = \frac{i}{2}\left(\frac{1}{\lambda} - \lambda\right) \\ \lambda &= e^{i\theta} \end{aligned} \right\} \tag{12-5}$$

we obtain by superposition the general solution

$$\varphi = \int_C F(X, \lambda)d\lambda \tag{12-6}$$

where C is a suitable contour in the λ plane.

With $X = x - y \cos \theta - z \sin \theta$, Eq. 12-2 is seen to be the potential of a plane wave at an angle of 45° to the x axis ($M_\infty = \sqrt{2}$) and at a variable angle θ to the negative z axis. Eq. 12-6 then represents the three-dimensional potential as the result of the superposition of plane waves whose form and intensity are functions of the angle θ.

Real values of θ correspond to points λ on the unit circle. On the other hand, for values of λ near the origin we have

$$X \stackrel{\rightarrow}{=} -\frac{1}{2\lambda}(y + iz) \tag{12-7}$$

and Eq. 12-2 represents a cylindrical, or two-dimensional, complex potential function which is a solution of

$$-\varphi_{yy} - \varphi_{zz} = 0$$

with $\varphi_{xx} = 0$ separately.

The form of Eq. 12-6 for the supersonic point source may be shown to be

$$\varphi_s = \frac{1}{4\pi^2} \int_C \frac{d\lambda}{i\lambda X} \tag{12-8}$$

To evaluate this integral by the method of residues it is necessary to determine the poles of the integrand. The denominator λX may be factored into

$$\lambda X = -\tfrac{1}{2}(y - iz)(\lambda - \epsilon_1)(\lambda - \epsilon_2) \tag{12-9}$$

where

$$\left.\begin{aligned} \epsilon_1 &= \frac{y + iz}{x + R} \\[2mm] \epsilon_2 &= \frac{y + iz}{x - R} \end{aligned}\right\} \tag{12-10}$$

and

$$R = \sqrt{x^2 - y^2 - z^2}$$

The integral in Eq. 12-8 may now be written

$$\varphi_s = -\frac{1}{4\pi^2} \int_C \frac{2d\lambda}{i(y - iz)(\lambda - \epsilon_1)(\lambda - \epsilon_2)} \tag{12-11}$$

Now consider the evaluation of Eq. 12-11 for a closed contour drawn just inside the unit circle on the λ plane. If the real part of R is taken as positive, then for values of x, y, z in the downstream branch of the Mach cone ($x^2 > y^2 + z^2$, $x > 0$), ϵ_1 will lie inside the unit circle while ϵ_2 lies outside (see Fig. A,12b). The value of the integral is then

$$\varphi_s = -\frac{1}{\pi(y - iz)(\epsilon_1 - \epsilon_2)} = \frac{1}{2\pi R} \tag{12-12}$$

For values of x, y, z in the upstream branch of the Mach cone, ϵ_1 and ϵ_2 are interchanged and ϵ_2 lies inside the unit circle. Hence we have

$$\varphi_s = -\frac{1}{\pi(y - iz)(\epsilon_2 - \epsilon_1)} = -\frac{1}{2\pi R} \tag{12-13}$$

Investigation of Eq. 12-10 shows that, for every point x, y, z in the "zone of silence" between the two branches of the Mach cone, both ϵ_1 and ϵ_2 lie on the unit circle and the value of the integral in Eq. 12-11 is then zero.

Exactly the same determination of φ results from a contour drawn in the opposite direction around the remaining portion of the λ surface outside the unit circle. It is found that integration and differentiation of the expression for the source are simplified if the complete contour C is speci-

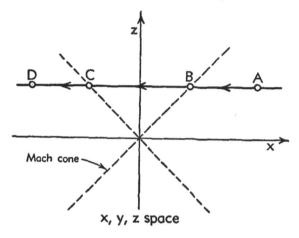

x, y, z space

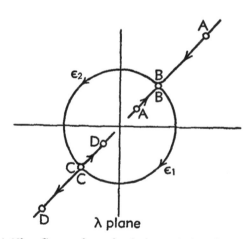

λ plane

Fig. A,12b.   Course of ϵ_1 and ϵ_2 during variation of point x, y, z.

fied so as to enclose the two portions of the λ surface $|\lambda| < 1$ and $|\lambda| > 1$ in opposite directions and to exclude the points $|\lambda| = 0$ and $|\lambda| = \infty$ (see Fig. A,12c).

The solution for a horseshoe vortex, which corresponds to the disturbance produced by an element of lift, may be obtained by the familiar process of integrating the expression for the source along x and then differ-

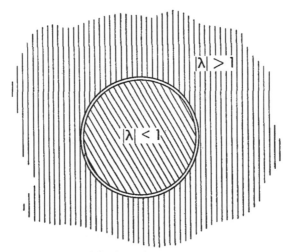

λ surface, showing cut

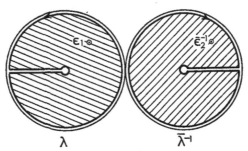

Fig. A,12c.   Representation of λ surface, showing two parts of contour C.

entiating with respect to z. The result is

$$\varphi_v = \frac{1}{4\pi^2} \int_C \frac{d\beta}{\alpha x - \beta y - \gamma z} = \begin{cases} 0 & \text{when } x^2 \leqq y^2 + z^2 \\ -\dfrac{xz}{\pi R(y^2 + z^2)} & \text{when } x > 0 \\ +\dfrac{xz}{\pi R(y^2 + z^2)} & \text{when } x < 0 \end{cases} \qquad (12\text{-}14)$$

The contour C, transformed according to the relation

$$\beta = \frac{1}{2}\left(\lambda + \frac{1}{\lambda}\right)$$

encloses the poles

$$\nu_1 = \frac{xy - iRz}{y^2 + z^2} = \frac{1}{2}\left(\epsilon_1 + \frac{1}{\epsilon_1}\right)$$

and

$$\nu_2 = \frac{xy + iRz}{y^2 + z^2} = \frac{1}{2}\left(\epsilon_2 + \frac{1}{\epsilon_2}\right)$$

in opposite directions, and excludes that part of the real axis between ± 1.

The combined disturbance field for an entire lifting surface S may now be obtained by superimposing elementary solutions of the form (12-14). This superposition amounts to an integration of elementary horseshoe vortices over the surface, the strength of the vortices at each point being determined by the local lift

$$l(x_1, y_1) = 2\rho_\infty U u(x_1, y_1, 0) \tag{12-15}$$

where $u(x_1, y_1, 0)$ denotes the value of the longitudinal perturbation velocity on the upper side of the lifting surface. After the order of integration has been changed so that the contour integration is performed last, the expression for the combined potential $\bar{\varphi}$ of the lifting surface becomes

$$\bar{\varphi} = \frac{1}{4\pi^2} \int_C \iint_S \frac{u(x_1, y_1, 0)\, dx_1 dy_1}{\alpha(x - x_1) - \beta(y - y_1) - \gamma z}\, d\beta \tag{12-16}$$

By differentiating Eq. 12-16 for the downwash velocity $\bar{w} = \partial\bar{\varphi}/\partial z$ there is obtained

$$\bar{w} = \frac{1}{4\pi^2} \int_C \iint_S \gamma \frac{u(x_1, y_1, 0)\, dx_1 dy_1}{[\alpha(x - x_1) - \beta(y - y_1) - \gamma z]^2}\, d\beta \tag{12-17}$$

As the result of the change in the order of integration, Eq. 12-16 and 12-17 now represent the three-dimensional flow as infinitely many two-dimensional fields superimposed. For each particular value of λ or β, the integration over the surface S yields an elementary two-dimensional disturbance field associated with an oblique strip drawn in the plane of the wing and having its edges tangent to the outline of the planform (see Fig. A,12d).

These relations are most easily seen for those parts of the contour C near $\lambda = 0$ and $1/\lambda = 0$. Considering first the loop around $\lambda = 0$, we have

$$\beta \xrightarrow{} \frac{1}{2\lambda}$$

$$\gamma \xrightarrow{} \frac{i}{2\lambda}$$

and

$$\alpha(x - x_1) - \beta(y - y_1) - \gamma z \xrightarrow{} \frac{1}{2\lambda}[y_1 - (y + iz)]$$

since $\alpha(x - x_1)$ is negligible by comparison. With these substitutions, Eq. 12-17 becomes,

$$\bar{w} = \frac{1}{4\pi^2} \int_C \iint_S \frac{u(x_1, y_1) \, dx_1 dy_1}{[y_1 - (y + iz)]^2} \frac{i}{\lambda} \, d\lambda \qquad (12\text{-}18)$$

The integration along x_1 may now be performed directly, and yields a

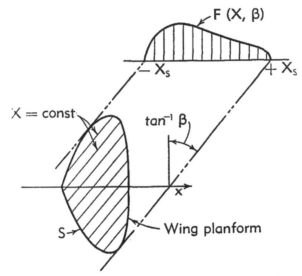

Fig. A,12d. Integration of surface loading along oblique lines $X = $ const.

quantity which is proportional to the spanwise distribution of circulation, i.e.

$$\int_S u(x_1, y_1) dx_1 = \tfrac{1}{2}\Gamma(y_1) \qquad (12\text{-}19)$$

The integral over the surface S reduces to

$$\int_S \frac{\tfrac{1}{2}\Gamma(y_1)}{[y_1 - (y + iz)]^2} \, dy_1 = \frac{1}{2} \int_S \frac{\Gamma'(y_1)}{y_1 - (y + iz)} \, dy_1 \qquad (12\text{-}20)$$

which is the familiar formula for the complex velocity function associated with the wake of trailing vortices.

Values of λ near $1/\lambda = 0$ yield an expression similar to Eq. 12-20 except that $y - iz$ appears instead of $y + iz$. The addition of the contributions of the two parts of the contour $|\lambda| \rightleftharpoons 0$ and $1/|\lambda| \rightleftharpoons 0$ thus yields a real value of $\bar{w}$ which corresponds to the downwash induced by the vortex wake. In particular, it is known that this component of the downwash will be constant if the integrated spanwise loading of the wing is elliptical.

Those portions of the contour C, extending around the unit circle in the λ plane or around the unit strip $(-1 < \beta < +1)$ in the β plane, yield a second component of $\bar{w}$ which corresponds to the "wave drag" of the lifting surface. Again, the surface integral in Eq. 12-17 may be shown to reduce to the expression for a two-dimensional velocity function. The form of the velocity function in this case depends, for each particular value of β, on an integration of the surface loading of the wing in an oblique direction.

To illustrate this relation we replace $\alpha x - \beta y - \gamma z$ by the single variable

$$X = \alpha x - \beta y - \gamma z$$

and for the variable point on the wing we introduce

$$X_1 = \alpha x_1 - \beta y_1 \tag{12-21}$$

together with the orthogonal variable

$$Y_1 = \frac{\beta x_1 + \alpha y_1}{\alpha^2 + \beta^2} \tag{12-22}$$

The factor $1/(\alpha^2 + \beta^2)$ in the latter expression preserves the elementary area. Eq. 12-17 may now be written

$$\bar{w} = \frac{1}{4\pi^2} \int_C \iint_S \frac{u(X_1, Y_1)dX_1dY_1}{(X - X_1)^2} \gamma d\beta \tag{12-23}$$

The integration over Y_1 may be performed directly and amounts simply to an integration of the loading of the wing along the oblique lines $X_1 = \text{const}$. Thus we may write

$$\int_S u(X_1, Y_1) \, dY_1 = F(X_1, \beta) \tag{12-24}$$

where $F(X_1, \beta)$ is the loading of the oblique lifting line, comparable to Γ in Eq. 12-19. The equation for $\bar{w}$ now becomes

$$\bar{w} = \frac{1}{4\pi^2} \int_C \int_{X-}^{X+} \frac{F(X_1, \beta) \, dX_1}{(X - X_1)^2} \gamma d\beta \tag{12-25}$$

The expression under the integral is the familiar formula for the downwash of a lifting line having the loading $F(X_1, \beta)$. Eq. 12-25 thus represents the downwash of the wing as the sum of values contributed by an infinite number of lifting lines.

Drag of a given distribution of lift. The final value of the wave drag will be obtained by integrating the product of the lift $l(x, y)$ and the downwash angle $\bar{w}/U$ over the surface S. In this way we obtain

$$D_w = - \rho_\infty \iint_S u\bar{w} \, dx dy \qquad (12\text{-}26)$$

where u corresponds to the specified distribution of lift and $\bar{w}$ is given by Eq. 12-25. Making use of Eq. 12-21, 12-22, and 12-24, we obtain

$$D_w = - \frac{\rho_\infty}{4\pi^2} \int_C \int_{X-}^{X+} \int_{X-}^{X+} \frac{F(X, \beta)F(X_1, \beta)dX dX_1}{(X - X_1)^2} \gamma d\beta \qquad (12\text{-}27)$$

Eq. 12-27 represents the wave drag of the lifting surface as the sum of the drags of the oblique lifting lines. Following a well-known procedure in subsonic wing theory, we may represent each of the oblique loadings of the wing by a Fourier series. Thus if

$$X + \text{const} = X_s \cos \delta$$

and

$$F(X, \beta) = \sum_1^\infty A_n \sin n\delta$$

the drag of each oblique lifting line will be proportional to $\sum_1^\infty nA_n^2$. Since the total lift L depends only on A_1, we obtain finally

$$D_w = \frac{1}{\pi} \int_{-1}^{+1} \frac{L^2}{\pi^{\frac{1}{2}}\rho_\infty U^2(2X_s)^2} \left(1 + \sum_2^\infty na_n^2\right) \sqrt{1 - \beta^2} \, d\beta \qquad (12\text{-}28)$$

where $a_n = A_n/A_1$ and the integration around the contour C has been replaced by four times the real integral over -1 to $+1$.

In Eq. 12-28 the quantity $2X_s$, which is a function of β, represents the "oblique span" of the lifting surface (see Fig. A,12d). For the particular value $\beta = 0$, $2X_s$ is simply the over-all length of the wing. The coefficients a_n, which will also generally vary with β, represent the departures of the loading curves from the elliptic form. As the formula shows, for a wing of given dimensions, departures from elliptic loadings have the effect of increasing the wave drag.

Eq. 12-28 is entirely analogous to the well-known formula [140] for the induced drag or vortex drag

$$D_v = \frac{L^2}{\pi^{\frac{1}{2}}\rho_\infty U^2 b^2} (1 + 2a_2^2 + 3a_3^2 + \cdots)$$

where the a_n's are now the coefficients occurring in the spanwise load distribution. It is evident that, in order to minimize both the wave drag and the vortex drag, the span loading and the oblique loadings for the range $-1 < \beta < 1$ should be as nearly elliptical as possible.

⟨ 143 ⟩

In the case of a long narrow wing, the preceding formulas may be simplified by the assumption that the projected dimension of the planform does not differ significantly from the over-all length c measured along the x axis. If we further assume that both the lengthwise and spanwise loadings are elliptical, so as to obtain the minimum drag, there results the formula

$$D_v + D_w = \frac{L^2}{\pi \frac{1}{2} \rho_\infty U^2 b^2} + \frac{M_\infty^2 - 1}{2} \frac{L^2}{\pi \frac{1}{2} \rho_\infty U^2 c^2} \qquad (12\text{-}29)$$

As pointed out by E. W. Graham, in certain cases the optimum distribution of lift will not cover the whole area of the allotted planform.

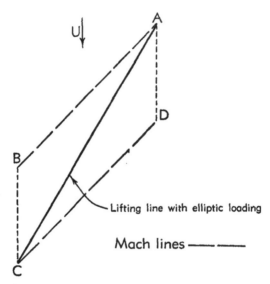

Fig. A,12e.   Ideal distribution of lift for parallelogram $ABCD$.

Thus if we describe a parallelogram bounded by Mach lines and lines parallel to the direction of flight, as shown in Fig. A,12e, the ideal distribution will place all the lift along a narrow lifting line extending diagonally across the area. Within the region shown in Fig. A,12f, formed by Mach-line and streamwise tangents to an ellipse, a uniform concentration of lift over the ellipse, with zero lift elsewhere, will result in the least drag.

The minimum drag of a wing having an elliptic planform. The foregoing discussion indicates that the minimum drag of a wing of elliptic planform will occur when the lift is distributed uniformly over the surface, since the integrated loading in every oblique direction is then elliptical.

The downwash $\bar{w}$ over the elliptic wing may be calculated directly from Eq. 12-17. To evaluate the surface integral we write

$$\iint\limits_{S} \frac{u(x_1,\, y_1)dx_1dy_1}{[\alpha(x - x_1) - \beta(y - y_1) - \gamma z]^2} = \pi i F'(X,\, \beta) \qquad (12\text{-}30)$$

After introducing $u = u_0$ and $X = \alpha x - \beta y - \gamma z$, and integrating over x_1, we obtain

$$F'(X,\, \beta) = \frac{1}{\pi i} \int_{-y_s}^{+y_s} \frac{1}{\alpha} \left[\frac{u_0 dy_1}{X - \alpha x_1 + \beta y_1} \right]_{-x_s}^{+x_s} \qquad (12\text{-}31)$$

where x_s and y_s correspond to the edges of the planform. For the ellipse

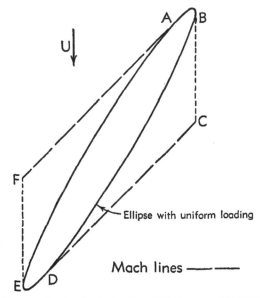

Fig. A,12f. Ideal distribution of lift for area $ABCDEF$.

with semiaxes a and b (Fig. A,12g) we have in parametric form

$$y_1 = b \cos \vartheta$$

$$x_s = -a \sin \vartheta$$

The integral in Eq. 12-31 now becomes

$$F'(X,\, \beta) = \frac{u_0 b}{\pi i a} \int_0^{2\pi} \frac{\sin \vartheta d\vartheta}{X + b\beta \cos \vartheta + a\alpha \sin \vartheta} \qquad (12\text{-}32)$$

The evaluation of integrals of this form is given in [53]. There are two determinations, depending on the location of the poles of the integrand

$$e_1 = \frac{-X - \sqrt{X^2 - (b\beta)^2 - (a\alpha)^2}}{b\beta - ia\alpha} \qquad (12\text{-}33)$$

and

$$e_2 = \frac{-X + \sqrt{X^2 - (b\beta)^2 - (a\alpha)^2}}{b\beta - ia\alpha} \qquad (12\text{-}34)$$

For points x, y inside the elliptic disk we have

$$\frac{x^2}{a^2} + \frac{y^2}{b^2} < 1$$

and in this case the general formula

$$\int_0^{2\pi} \frac{\cos n\vartheta\, d\vartheta}{X + b\beta \cos \vartheta + a\alpha \sin \vartheta} = \int_0^{2\pi} \frac{i \sin n\vartheta\, d\vartheta}{X + b\beta \cos \vartheta + a\alpha \sin \vartheta}$$

$$= \pi \frac{e_1^n - e_2^n}{\sqrt{X^2 - (b\beta)^2 - (a\alpha)^2}} \qquad (12\text{-}35)$$

applies. Eq. 12-32 corresponds to $n = 1$, and the formulas give

$$F'(X, \beta) = \frac{2u_0 b}{\alpha} \frac{1}{b\beta - ia\alpha} \qquad (12\text{-}36)$$

Higher values of n correspond to nonuniform loadings over the wing. It is evident that in the present case $F'(X, \beta)$ is independent of X over

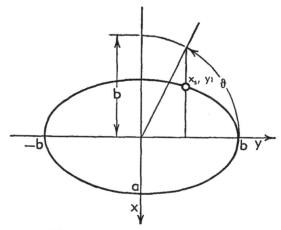

Fig. A,12g. Notation for elliptic wing.

the surface of the ellipse and hence the downwash will be constant at these points.

The value of the downwash is now obtained by introducing the value (12-36) into (12-30) and substituting in Eq. 12-17, making use of the relations

$$\alpha = 1$$
$$\gamma = \sqrt{1 - \beta^2}$$

Thus

$$\frac{\bar{w}}{U} = \frac{u_0}{2\pi i U} \int_C \frac{\sqrt{1-\beta^2}}{\beta - i(a/b)} = \frac{2u_0}{U}\sqrt{1 + \frac{a^2}{b^2}} \qquad (12\text{-}37)$$

Since the value $\beta = i(a/b)$ corresponds to two distinct poles, one on each portion of the λ surface (or, one pole on each sheet of the β surface), the value of the integral is twice the residue at this point.

The drag is given by

$$D_{\min} = \frac{1}{2}\frac{\bar{w}}{U}L$$

where L is the total lift, equal to

$$L = 2\rho_\infty u_0 U S$$

for a wing of area S. The formula for the minimum drag of the elliptic wing then becomes

$$D_{\min} = \frac{L^2}{4(\frac{1}{2}\rho_\infty U^2)S}\sqrt{1 + \frac{a^2}{b^2}} \qquad (12\text{-}38)$$

or, in terms of the coefficients C_L and C_D,

$$C_{D_{\min}} = C_L^2 \sqrt{\left(\frac{1}{4}\right)^2 + \left(\frac{1}{\pi R}\right)^2} \qquad (12\text{-}39)$$

In the latter form it can be seen that the first term under the radical sign corresponds to the value given by the Ackeret theory for a wing of very high aspect ratio, while the second term corresponds to the value given by the Munk theory. The latter value is approached at small values of the aspect ratio R, for which the vortex drag predominates over the wave drag.

Eq. 12-37 and 12-39 apply at $M_\infty = \sqrt{2}$. The variation with Mach number may be deduced, however, by correcting the downwash $\bar{w}/U$ by the factor $\sqrt{M_\infty^2 - 1}$ and multiplying the ratio a/b by the factor $1/\sqrt{M_\infty^2 - 1}$.

For the elliptic wing in straight flow, the drag is invariably greater than that given by the Ackeret theory for a two-dimensional flat plate. It is to be expected that lower values could be obtained if a long narrow ellipse were yawed behind the Mach cone.

The treatment of the yawed ellipse follows the preceding analysis closely. Again the minimum drag occurs when the lift density is constant over the whole surface. If the equation of the ellipse is written in the form

$$x_e = m'y_1 \pm a'\sqrt{1 - \frac{y_1^2}{b'^2}}$$

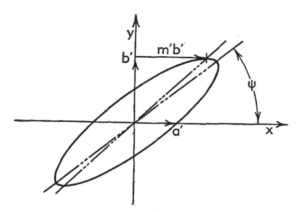

Fig. A,12h. Oblique ellipse.

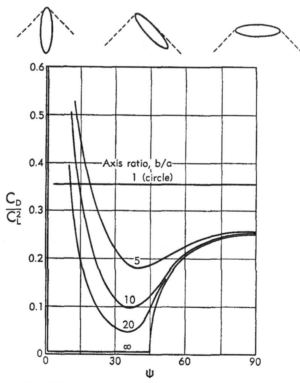

Fig. A,12i. Minimum drag of elliptic wings at various angles of yaw.

(see Fig. A,12h), the expression for the drag reduces to

$$D_{\min} = \frac{L^2}{4(\frac{1}{2}\rho_\infty U^2)S} \text{ R.P. } \sqrt{1 - \left(m' + i\frac{a'}{b'}\right)^2} \qquad (12\text{-}40)$$

The limit $a'/b' \to 0$ corresponds to infinite aspect ratio and the expression for the drag then involves

$$\text{R.P. } \sqrt{1 - m'^2}$$

The change of this quantity from real to imaginary, as m' passes through 1, shows the disappearance of the wave drag when the wing of infinite aspect ratio is yawed behind the Mach cone. The variation of drag coefficient with angle of yaw ψ for ellipses of finite proportions is shown in Fig. A,12i.

A,13. Determination of Lift in Three-Dimensional Flow. Methods and Formulas.

INTRODUCTION. The preceding article describes a general method for finding the drag when the distribution of lift on a surface is given. Finding the shape of the surface under the same circumstances is generally a direct mathematical procedure; this problem is adequately discussed in VI,D,11. The problem of finding the lift distribution when the wing geometry is given is more difficult. We are first of all confronted with the problem of mixed boundary conditions, the pressure being specified off the wing surface and the vertical component of velocity on it. Consequently the known relations take the form of an integral equation, solution of which is possible only under certain limited conditions. The solutions are not determinate without a knowledge of the singularities. This latter information is obtained from physical considerations and is based on the previously discussed classification of the edges (Art. 10). It is obvious that this classification is not so much a matter of the wing planform as it is of the speed range. Every sweptback wing, for example, on entering the supersonic regime, has subsonic leading edges, and in most cases, subsonic trailing edges. At a higher Mach number the same planform may have subsonic leading edges and supersonic trailing edges. Finally, if the Mach number is increased sufficiently, both leading and trailing edges will be supersonic. Since the interference effects also vary radically with the flight speed because of the dependence of the extent of the disturbance fields on the Mach angle, it is clear that no single concise formula can describe the flow over a given wing throughout the supersonic speed range.

Nevertheless, we are in a far better position with regard to the lifting surface problem at supersonic speeds than we were (Art. 7) with the wing at subsonic speeds. Just because the interference effects (in the first order theory) are confined to the interior of the Mach cones from the various points of disturbance, it is possible to perform far more complete calculations of the flow over most planforms than in the corresponding subsonic problems.[12] It is no longer necessary to resort to the device of the lifting line approximation, and for many planforms the entire lifting surface solution can be obtained with no simplifying assumptions beyond those used in linearizing the flow equation.

The methods presently used fall into two broad classifications which may be termed, from their elements, conical flow methods and source-distribution methods. The choice between them depends largely on the planform and Mach number under consideration. In many cases, it will be more effective to treat part of the problem by one method and part by the other, than to use either alone [141]. When the planform is a simple polygon, the major part of the flow field can usually be described immediately in terms of one or more conical flow solutions. For this reason the conical flow method is presented first.

CONICAL FLOW THEORY. As originally conceived by Busemann [57], a conical flow field is defined as a field in which all the velocity components, and hence the pressure and the slope of the streamlines, are constant along rays emanating from a single point. An application of conical solutions in subsonic flow has been described in Art. 7 and another indicated in Art. 9. However, the chief use of conical disturbance fields has been in connection with supersonic flow, where the Mach cone bounding the field of a point disturbance is itself one of the conical surfaces of constant pressure of the field. Busemann found that by restricting the problem to small disturbances, solutions of very general and elegant form could be obtained. We further confine ourselves here to essentially planar disturbances. For these, the work of Lagerstrom [142], Hayes [143], Germain [144], and others has made available an extensive and useful theory, of which we will present, with a minimum of development, only the most directly applicable results.

Solutions of conical form constitute part of a more general class of theoretical flows having considerable interest from the standpoint of practical wing shapes. These flows are described by homogeneous functions of the coordinates x, y, and z, and contain the conical fields as the special class of degree zero. Solutions of higher degree are sometimes called "quasi-conical," and the pressure, instead of being constant, varies according to some power of x, y, or z along each ray. An example of such a field has already been encountered (Art. 9) through the integration of the conical field of a simple wedge to obtain a parabolically curved surface. Another method of obtaining solutions of quasi-conical form is through application of the easily proved theorem that, if u, v, and w are the velocity components of any disturbance field satisfying the linearized flow

[12] The wing of elliptic planform must be noted as an exception to this statement.

equation, then

$$f = xu + yv + zw$$

also satisfies the equation. If u, v, and w are components of a conical velocity field—or more generally, are homogeneous functions—the solution f will obviously be of one higher degree in the coordinates.

The general subject of quasi-conical fields has been treated by Heaslet and Lomax in this series (VI,D,13), Germain [144], Robinson [145], Squire [146], and Multhopp [147]. Applications to rolling and pitching wings may be found in [148,149,150]. The reader will have no difficulty in generalizing much of what follows to similar problems in quasi-conical flow.

Returning to the consideration of the conical solutions, comparison of the three-dimensional Laplace equation and the linearized equation for supersonic speeds in the normalized form (10-4) suggests that solutions of the latter may be obtained by replacing y by $\pm iy$ and z by $\pm iz$ in the general solution (7-27) of the former. The principal result is the replacement of the radial distance $\sqrt{x^2 + y^2 + z^2}$ by the so-called hyperbolic distance $\sqrt{x^2 - y^2 - z^2}$. The numerator of the argument becomes $\mp(z \pm iy)$, but because y and z are interchangeable in Eq. 10-4, we may choose a form for our variable more closely resembling the subsonic one, namely

$$\epsilon = \frac{y + iz}{x + \sqrt{x^2 - y^2 - z^2}} \tag{13-1}$$

It can be shown [142] that Eq. 10-4, written in terms of the real and imaginary parts of ϵ, reduces identically to the two-dimensional Laplace equation. However, since the new coordinates bear no direct relation to the velocity components (such as Eq. 2-7), this fact does not offer as great an advantage as might be expected. Instead it is convenient in the supersonic case to make use of a further transformation of variables, writing

$$\tau = \frac{2\epsilon}{1 + \epsilon^2} \tag{13-2}$$

The variable τ has the property of mapping the interior of the Mach cone into the whole of the complex plane. The Mach cone itself and the region outside it are transformed by Eq. 13-2 into the positive and negative branches of the real axis outside ± 1 and the circle at infinity (see Fig. A,13a). In the neighborhood of the x, y plane, the variable τ approaches the value $(y + iz)/x$, so that the contour of a thin airfoil is mapped undistorted into the neighborhood of the real axis of the τ plane. Thus the expression of the boundary conditions in terms of τ is very simple.

The relations between the components of the velocity, obtained from

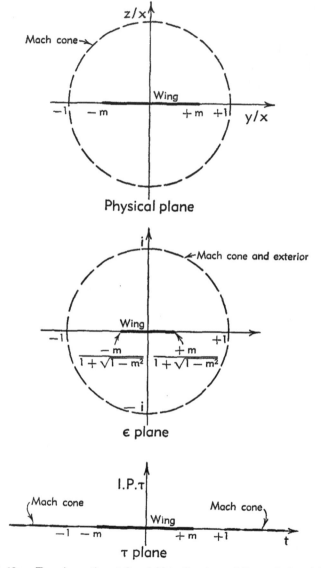

Fig. A,13a. Transformation of flow field to the plane of the conical variable τ.

the irrotationality conditions, are

$$d\hat{W} = -i \frac{\sqrt{1 - \tau^2}}{\tau} d\hat{U} \tag{13-3}$$

and

$$d\hat{V} = -\frac{1}{\tau} d\hat{U} \tag{13-4}$$

where $\hat{U}$, $\hat{V}$, and $\hat{W}$ are the analytic functions of τ of which u, v, and w are the real parts.

If the given boundary conditions are sufficient to determine either $\hat{U}$ or $\hat{W}$, the remaining components can be found by integrating. The conditions to be satisfied on the Mach cone determine the constant of integration, except for an imaginary part of no significance in the final application. When the boundary conditions are "mixed," that is $\hat{U}$ is conditioned in some regions and $\hat{W}$ in others, Eq. 13-3 yields some general conclusions useful in translating the conditions on $\hat{W}$ into terms of $\hat{U}$. First, it is seen that, in the plane of the wing ($z = 0$) and within the Mach cone ($|\tau| < 1$), a necessary condition for a flat airfoil surface, $w = $ const, is that the imaginary part of $\hat{U}$ be zero. However, this condition is not sufficient, for if $\hat{U}' = d\hat{U}/d\tau$ is continuous in the neighborhood of $\tau = 0$, integrating across the singularity in Eq. 13-3 will result in a discontinuity in $\hat{W}$ of the magnitude $-\pi \hat{U}'(0)$, corresponding to a step in the downwash, unless R.P. $\hat{U}'(0)$ is equal to zero.[13] Thus for a continuous surface of asymmetric planform including the $\tau = 0$ axis, two terms are generally required to express the pressure distribution, so that the lateral gradient may be zero through the axis.

We are further aided in the construction of solutions by the following considerations concerning the flow at the edges:

1. If the edge is a leading edge at an angle of attack, and the flow normal to it is subsonic, there will be a $\frac{1}{2}$-power singularity in the pressure on the wing and in the downwash just ahead of the leading edge.
2. If the normal component of the stream velocity is supersonic, both u and w will be stepwise discontinuous on crossing the edge.
3. If the edge is trailing and the normal flow is subsonic, application of the Kutta condition determines that the discontinuity in u at the plane of the wing will vanish at the edge, and that w will remain finite on crossing the edge.
4. If the edge is trailing, but the normal flow is supersonic, the Kutta condition may be replaced by a discontinuous expansion to zero pressure. (Nonlifting wings are discussed elsewhere.)

[13] If I.P. $\hat{U}'(0)$ is not equal to zero there will be a logarithmic infinity in the downwash along the ray $\tau = 0$.

The logarithmic functions have been found most useful in setting up solutions to fit simple boundary conditions. Fig. A,13b illustrates the behavior of these functions. If $\zeta(\tau)$ is any complex function of τ which on the plane of the wing takes on real values, then the real part of $i/\pi \ln \zeta(t)$ will provide a unit step wherever $\zeta(t)$ goes through zero. The function $\zeta(t)$ may be otherwise constructed to cause the disturbance field to vanish at infinity ($\tau \rightarrow \infty$) and to satisfy the edge conditions where desired.

Conversely, the inverse trigonometric functions (see Fig. A,13b) can be set up to satisfy the Kutta condition at a trailing edge and provide

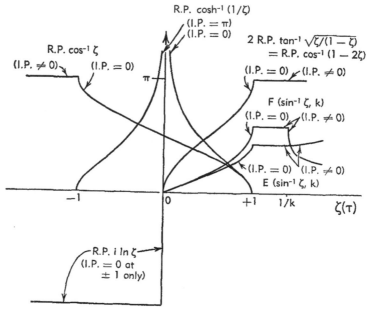

Fig. A,13b.   Sketch showing real parts of functions useful as conical flow solutions.

constant pressure (or downwash) over some other region of the field; it will follow that in the region in which the real part of the function varies, the downwash (or pressure) from Eq. 13-3, will either be constant or step-shaped. The elliptic integrals $F(\vartheta, k)$ and $E(\vartheta, k)$ show essentially the same pattern, as sin ϑ varies from zero, through one, to $1/k$ [151]. Square-root terms will be needed to provide the singularities at leading edges and also give constant or step-shaped downwash or pressure curves. The inverse hyperbolic functions result from assuming one velocity component in logarithmic form.

A table of conical flow solutions[14] follows with sketches of the bound-

[14] Most of these solutions are derived in [142]. A few solutions have been derived especially for this table. References for the others are as follows: solutions 10 and 11, [152]; solution 12, [146]; solution 13, [148]; solutions 13 and 14, [153].

ary conditions they are designed to satisfy. Only the u and w components of the flows are given, but since these are given in complex form, the sidewash v, if desired, can be obtained by the use of Eq. 13-4. The velocities are plotted against the (real) values of τ in the plane $z = 0$. The symbol m usually denotes the value of τ along the leading or trailing edge of the wing, that is, $\sqrt{M_\infty^2 - 1}$ times the inclination of the edge to the stream. However, it may merely refer to the ray separating two regions where different boundary conditions apply, and in the superposition procedures to be described later, m will often be replaced by a variable.

The double signs in the designation of u and w refer to values on the upper and lower surfaces and indicate discontinuities in their values across the plane of the wing. A discontinuity in w indicates divergence of the upper and lower surfaces of the wing, or thickness; a discontinuity in u corresponds to lift. However, when there is no possibility of interaction between the upper and lower surfaces, as in the case of a wing with supersonic edges, or of a control surface at the rear of a wing, any thickness solution may be used to obtain the lift on a thin surface of similar inclination by reversing the signs of the solution to take account of the actual slopes of the upper and lower surfaces (cf. solutions 2 and 3). The lifting solutions may be used in the same way to determine the thickness distribution corresponding to a specified pressure distribution.

Table A,13a is of course not exhaustive; other solutions are to be found in [*142,152*] and elsewhere. On the other hand, several of the solutions presented are readily derivable one from another, but are included because of their individual usefulness or to indicate methods of devising additional solutions. For example, the solution (2) for the bilaterally symmetrical nonlifting wedge is obtained from the solution (1) for the semi-infinite wedge by the superposition of the latter and its reflection in the x axis. Solutions 2 and 10 are related in that the vertical component in the former is required to satisfy the same boundary conditions as the streamwise component in the latter, so that the same function will do for both, the remaining components in each case being determined from Eq. 13-3 and 13-4. Some of the solutions are simply special cases of others, but conversely some may be regarded as generalizations of others, obtainable (e.g. 7 from 6) by an oblique transformation, discussed immediately below. It should be possible to satisfy any simple combination of conical boundary conditions with a modified form of the solutions given.

THE OBLIQUE TRANSFORMATION APPLIED TO CONICAL FLOWS. It may be helpful to discuss briefly the subject of the oblique transformation. The subsonic form of this transformation was used (except for a scale factor which has no significance in a conical field) in Art. 9 (Eq. 9-9) and corresponds to a simple rotation of the axis. Again omitting the scale

Table A,13a. Conical flow functions.

Flow field; u and w components	Values in plane of wing	Plan view and boundary conditions

1. Wedge, supersonic leading edge, streamwise side edge

$$u = \text{R.P.} \; \frac{-m}{\pi\sqrt{m^2-1}}\cos^{-1}\frac{1-m\tau}{m-\tau}$$

$$w = \text{R.P.} \; \frac{i}{\pi}\ln\frac{\tau}{\tau-m}$$

2. Unyawed wing, wedge section, $m > 1$

$$u = \text{R.P.} \; \frac{-m}{\pi\sqrt{m^2-1}}\left(\cos^{-1}\frac{1-m\tau}{m-\tau} + \cos^{-1}\frac{1+m\tau}{m+\tau}\right)$$

$$w = \text{R.P.} \; \frac{i}{\pi}\ln\frac{\tau+m}{\tau-m}$$

3. Flat lifting triangle, $m > 1$

$$u = \pm\text{R.P.} \; \frac{\alpha U}{\beta\pi}\frac{m}{\sqrt{m^2-1}}\left(\cos^{-1}\frac{1-m\tau}{m-\tau} + \cos^{-1}\frac{1+m\tau}{m+\tau}\right)$$

$$w = \mp\text{R.P.} \; i\frac{\alpha U}{\beta\pi}\ln\frac{\tau+m}{\tau-m}, \qquad \beta = \sqrt{M_\infty^2-1}$$

(Two independent flows of form 2)

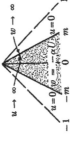

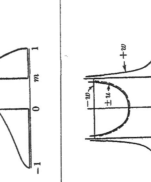

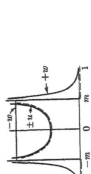

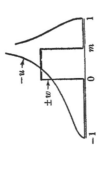

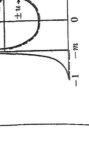

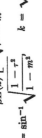

4. Yawed triangle, both leading edges supersonic

$$u = \pm \text{R.P.} \, \frac{\alpha U}{\beta \pi} \left(\frac{m_1}{\sqrt{m_1^2 - 1}} \cos^{-1} \frac{1 - m_1 \tau}{m_1 - \tau} + \frac{m_2}{\sqrt{m_2^2 - 1}} \cos^{-1} \frac{1 + m_2 \tau}{m_2 + \tau} \right)$$

$$w = \mp \text{R.P.} \, i \, \frac{\alpha U}{\beta \pi} \ln \frac{\tau + m_2}{\tau - m_1}$$

5. Wedge, subsonic leading edge, streamwise side edge

$$u = \text{R.P.} \, \frac{-m}{\pi \sqrt{1 - m^2}} \cosh^{-1} \frac{1 - m\tau}{m - \tau}$$

$$w = \text{R.P.} \, \frac{i}{\pi} \ln \frac{\tau}{\tau - m}$$

6. Flat lifting triangle, $m < 1$

$$u = \text{R.P.} \, \frac{m^2 \alpha U}{\beta E(k)} \, \frac{1}{\sqrt{m^2 - \tau^2}}$$

$$w = \text{R.P.} \, \frac{\alpha U}{\beta E(k)} \left[\tau \sqrt{\frac{1 - \tau^2}{\tau^2 - m^2}} - E(\vartheta, k) \right]$$

$$\vartheta = \sin^{-1} \sqrt{\frac{1 - \tau^2}{1 - m^2}}, \quad k = \sqrt{1 - m^2}$$

Flow field; u and w components	Values in plane of wing	Plan view and boundary conditions

7. Yawed triangle, both leading edges subsonic

$$u = \text{R.P.} \frac{\alpha U}{\beta A E(k)} \left[m_1 \sqrt{\frac{m_2 + \tau}{m_1 - \tau}} + m_2 \sqrt{\frac{m_1 - \tau}{m_2 + \tau}} \right]$$

$$w = \text{R.P.} \frac{\alpha U}{\beta A E(k)} \left\{ (1 + m_1) \sqrt{\frac{(\tau + m_2)(1 - \tau)}{(\tau - m_1)(1 + \tau)}} \right.$$

$$+ (1 - m_2) \sqrt{\frac{(\tau - m_1)(1 - \tau)}{(\tau + m_2)(1 + \tau)}}$$

$$\left. + A[(1 - \sqrt{1 - k^2 \sin^2 \vartheta})\cot\vartheta - E(\vartheta, k)] \right\}$$

$$A = \sqrt{(1 + m_1)(1 + m_2)} + \sqrt{(1 - m_1)(1 - m_2)}$$

$$k = \frac{2}{A}[(1 - m_1^2)(1 - m_2^2)]^{\frac{1}{4}},$$

$$\vartheta = \tan^{-1} \frac{A}{2} \sqrt{\frac{1 - \tau^2}{(\tau - m_1)(\tau + m_2)}}$$

8. Yawed triangle, one edge subsonic, one supersonic

$$u = \text{R.P.} \frac{2U\alpha}{\pi\beta} \left[\frac{m_1}{1 + m_1} \sqrt{\frac{(m_1 + m_2)(1 + \tau)}{(1 + m_2)(m_1 - \tau)}} \right.$$

$$\left. + \frac{m_2}{\sqrt{m_2^2 - 1}} \tan^{-1} \sqrt{\frac{(m_2 - 1)(m_1 - \tau)}{(m_1 + m_2)(1 + \tau)}} \right]$$

$$w = \text{R.P.} \frac{2U\alpha}{\pi\beta} \left[\sqrt{\frac{(m_1 + m_2)(1 - \tau)}{(1 + m_2)(\tau - m_1)}} \right.$$

$$\left. - \tan^{-1} \sqrt{\frac{(m_1 + m_2)(1 - \tau)}{(1 + m_2)(\tau - m_1)}} \right]$$

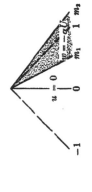

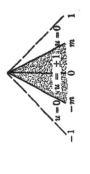

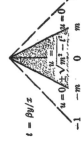

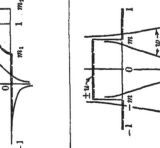

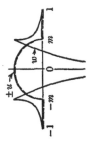

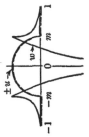

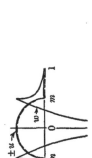

$l = \beta y/x$

9. Yawed triangle, one edge trailing

$$u = \text{R.P.}\ \frac{2m_2\alpha U}{\pi\beta\sqrt{m_1^2-1}}\tan^{-1}\sqrt{\frac{(m_2-1)(\tau-m_1)}{(m_2-m_1)(1-\tau)}}$$

$$w = \text{R.P.}\ \frac{2\alpha U}{\pi\beta\sqrt{m_2+1}}\left[\sqrt{\frac{m_2-m_1}{m_1}}\right.$$

$$\times \sinh^{-1}\sqrt{\frac{m_1-\tau}{(1+m_1)\tau}}$$

$$\left.-\sqrt{m_2+1}\tan^{-1}\sqrt{\frac{(m_2-m_1)(1+\tau)}{(m_2+1)(m_1-\tau)}}\right]$$

10. Uniformly loaded triangular wing

$$u = \text{R.P.}\ \frac{i}{\pi}\frac{1}{\tau}\ln\frac{\tau+m}{\tau-m}$$

$$w = \text{R.P.}\ \frac{1}{\pi}\left[\frac{\sqrt{1-m^2}}{m}\left(\cosh^{-1}\frac{1+m\tau}{\tau+m}\right.\right.$$

$$\left.\left.+\cosh^{-1}\frac{1-m\tau}{\tau-m}\right)-\frac{2}{m}\cosh^{-1}\frac{1}{\tau}\right]$$

11. Elliptically loaded triangular wing

$$u = \text{R.P.}\ i(\tau-\sqrt{\tau^2-m^2})$$

$$w = \text{R.P.}\left[\sqrt{1-m^2}\sqrt{\tau^2-m^2}-\cosh^{-1}\frac{1}{\tau}\right.$$

$$\left.+F(\vartheta,k)-E(\vartheta,k)\right]$$

$$k = \sqrt{1-m^2},\qquad \vartheta = \sin^{-1}\sqrt{\frac{1-\tau^2}{1-m^2}}$$

Table A,13a (continued)

Flow field; u and w components	Values in plane of wing	Plan view and boundary conditions
12. Cone with elliptic cross section $u = \text{R.P.}\ \dfrac{-m^2}{1-m^3}[\tau\tan\vartheta + F(\vartheta,k) - E(\vartheta,k)]$ $k = \sqrt{1-m^2}, \qquad \vartheta = \sin^{-1}\sqrt{\dfrac{1-\tau^2}{1-m^2}}$ $w = \text{R.P.}\ \dfrac{-m^2}{\sqrt{m^2-\tau^2}}$		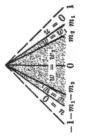 $t = \beta y/x$
13. Symmetrical flap deflection at leading edge $u = \text{R.P. }\cosh^{-1}\sqrt{\dfrac{m_1^2-m_2^2}{\tau^2-m_2^2}}$ $w = \text{R.P. }\sqrt{m_1^2-m_2^2}\,[\Pi(\vartheta,n,k) - F(\vartheta,k)]$ $w_1 = \sqrt{m_1^2-m_2^2}\left[\Pi\left(\dfrac{\pi}{2},n,k\right) - K(k)\right]$ $w_2 = w_1 - \dfrac{\pi}{2m_2}\sqrt{1-m_2^2}$ $\vartheta = \sin^{-1}\sqrt{\dfrac{1-\tau^2}{1-m_1^2}}, \qquad n = -\dfrac{1-m_1^2}{1-m_2^2}$ $k = \sqrt{1-m_1^2},$ $\Pi = \int_0^\vartheta \dfrac{d\theta}{(1+n\sin^2\theta)\sqrt{1-k^2\sin^2\theta}}$		

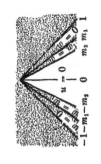

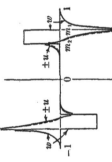

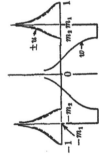

14. Antisymmetrical flap deflection at leading edge

$$u = \text{R.P.} \; \frac{2w_1}{\pi \sqrt{1-m_2^2}} \left[m_2 \cosh^{-1} \frac{m_2}{m_1} \frac{\sqrt{m_1^2-\tau^2}}{\sqrt{m_2^2-\tau^2}} - \tau \sqrt{\frac{m_1^2-m_2^2}{m_1^2-\tau^2}} \right]$$

$$w = \text{R.P.} \; \frac{2w_1}{\pi} \left[\tan^{-1} \sqrt{\frac{(m_1^2-m_2^2)(1-\tau^2)}{(1-m_2^2)(\tau^2-m_1^2)}} - \sqrt{\frac{(m_1^2-m_2^2)(1-\tau^2)}{(1-m_2^2)(\tau^2-m_1^2)}} \right]$$

15. Symmetrical flap deflection at trailing edge

$$u = \text{R.P.} \left[m_1 \sqrt{\frac{m_1^2-m_2^2}{1-m_1^2}} \Pi(\vartheta, n, k) + Z(\psi, k) F(\vartheta, k) \right]$$

$$w = \text{R.P.} \left\{ \frac{1}{m_2} \left[-Z(\psi, k) \right] + \frac{1}{m_1} \sqrt{(1-m_1^2)(m_1^2-m_2^2)} \; \sinh^{-1} \sqrt{\frac{m_2^2-\tau^2}{\tau^2}} \right. $$

$$\left. - \frac{\sqrt{1-m_1^2}}{m_1} \tan^{-1} \sqrt{\frac{m_1^2-m_2^2}{m_2^2-\tau^2}} \right\}$$

See 13 for Π

$$\vartheta = \sin^{-1}\sqrt{\frac{1-\tau^2}{1-m_2^2}}, \quad n = -\frac{1-m_2^2}{1-m_1^2}, \quad w_1 = -\frac{\pi \sqrt{1-m_1^2}}{2m_1}$$

$$k = \sqrt{1-m_2^2}, \quad \psi = \sin^{-1}\sqrt{\frac{1}{-n}}$$

$$Z(\psi, k) = E(\psi, k) - \frac{E(k) F(\psi, k)}{K(k)}$$

Table A,18a (continued)

Flow field; u and w components	Values in plane of wing	Plan view and boundary conditions

16. Antisymmetrical flap deflection at trailing edge

$$u = \text{R.P. } \cosh^{-1}\sqrt{\frac{(1-m_1^2)(\tau^2-m_2^2)}{(1-m_2^2)(\tau^2-m_1^2)}}$$

$$w = \text{R.P. } \frac{\sqrt{1-m_1^2}}{m_1}\tan^{-1}\frac{\tau}{m_1}\sqrt{\frac{m_1^2-m_2^2}{m_2^2-\tau^2}}$$

$$w_1 = \frac{\pi}{2}\frac{\sqrt{1-m_1^2}}{2m_1}$$

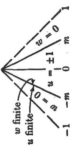

17. Side edge, wing with supersonic leading edge

$$u = \text{R.P. } \frac{2m\alpha U}{\beta\pi}\frac{1}{\sqrt{m^2-1}}\tan^{-1}\sqrt{\frac{-(1+m)\tau}{m(1+\tau)}}$$

$$w = \text{R.P. } \frac{2\alpha}{\beta\pi}U\left[\sqrt{\frac{m(1-\tau)}{\tau(m-1)}}-\tan^{-1}\sqrt{\frac{m(1-\tau)}{\tau(m-1)}}\right], \quad m>1$$

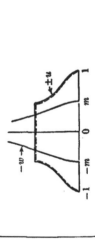

18. Symmetrical lift-cancellation field

$$u = \text{R.P. } \pm\frac{F(\vartheta,k)}{K(k)}$$

$$w = \text{R.P. } \frac{+1}{mK(k)}\cosh^{-1}\frac{m}{\tau}$$

$$\vartheta = \sin^{-1}\sqrt{\frac{1-\tau^2}{1-m^2}}, \quad k = \sqrt{1-m^2}$$

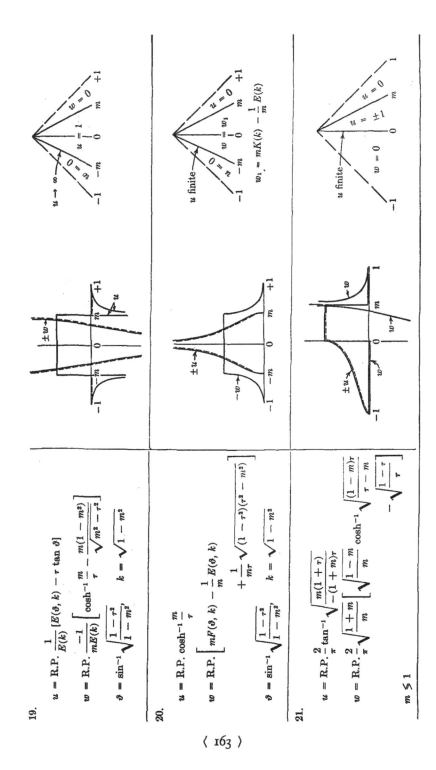

19.

$$u = \text{R.P. } \frac{1}{E(k)} \left[E(\vartheta, k) - \tau \tan \vartheta \right]$$

$$w = \text{R.P. } \frac{-1}{mE(k)} \left[\cosh^{-1} \frac{m}{\tau} - \frac{m(1 - m^2)}{\sqrt{m^2 - \tau^2}} \right]$$

$$\vartheta = \sin^{-1} \sqrt{\frac{1 - \tau^2}{1 - m^2}}, \qquad k = \sqrt{1 - m^2}$$

20.

$$u = \text{R.P. } \cosh^{-1} \frac{m}{\tau}$$

$$w = \text{R.P. } \left[mF(\vartheta, k) - \frac{1}{m} E(\vartheta, k) + \frac{1}{m\tau} \sqrt{(1 - \tau^2)(\tau^2 - m^2)} \right]$$

$$\vartheta = \sin^{-1} \sqrt{\frac{1 - \tau^2}{1 - m^2}}, \qquad k = \sqrt{1 - m^2}$$

21.

$$u = \text{R.P. } \frac{2}{\pi} \tan^{-1} \sqrt{\frac{m(1 + \tau)}{-(1 + m)\tau}}$$

$$w = \text{R.P. } \frac{2}{\pi} \sqrt{\frac{1 + m}{m}} \left[\sqrt{\frac{1 - m}{m}} \cosh^{-1} \sqrt{\frac{(1 - m)\tau}{\tau - m}} - \sqrt{\frac{1 - \tau}{\tau}} \right]$$

$$m \lessgtr 1$$

Flow field; u and w components	Values in plane of wing	Plan view and boundary conditions
22. $u = \text{R.P.} \dfrac{2}{\pi} \tan^{-1} \sqrt{\dfrac{(m_2 - m_1)(1 - \tau)}{(1 - m_1)(\tau - m_2)}}$ $w = \text{R.P.} \dfrac{2}{\pi} \sqrt{1 - m_1} \left[\sqrt{\dfrac{m_2 - m_1}{m_2}} \right.$ $\times \cosh^{-1} \sqrt{\dfrac{m_2 - \tau}{-(1 + m_2)\tau}}$ $\left. - \sqrt{1 + m_1} \cosh^{-1} \sqrt{\dfrac{(1 + m_1)(m_2 - \tau)}{(1 + m_2)(m_1 - \tau)}} \right]$		
23. $u = \text{R.P.} \dfrac{i}{\pi} \ln \dfrac{m_1 - \tau}{m_2 - \tau}$ $w = \dfrac{1}{\pi} \text{R.P.} \left[\dfrac{m_2 - m_1}{m_1 m_2} \cosh^{-1} \dfrac{1}{\tau} \right.$ $- \dfrac{\sqrt{1 - m_1^2}}{m_1} \cosh^{-1} \dfrac{1 - m_1\tau}{m_1 - \tau}$ $\left. - \dfrac{2}{m_2} \sqrt{m_2^2 - 1} \tan^{-1} \sqrt{\dfrac{(m_2 - 1)(1 + \tau)}{(m_2 + 1)(1 - \tau)}} \right]$		
24. $u = \dfrac{2m^{\frac{1}{2}}}{\pi(1 + m)} \text{R.P.} \sqrt{\dfrac{1 + \tau}{m - \tau}}$ $w = \dfrac{-2}{\pi} \text{R.P.} \left[\tan^{-1} \sqrt{\dfrac{m(1 - \tau)}{\tau - m}} - \sqrt{\dfrac{m(1 - \tau)}{\tau - m}} \right]$		

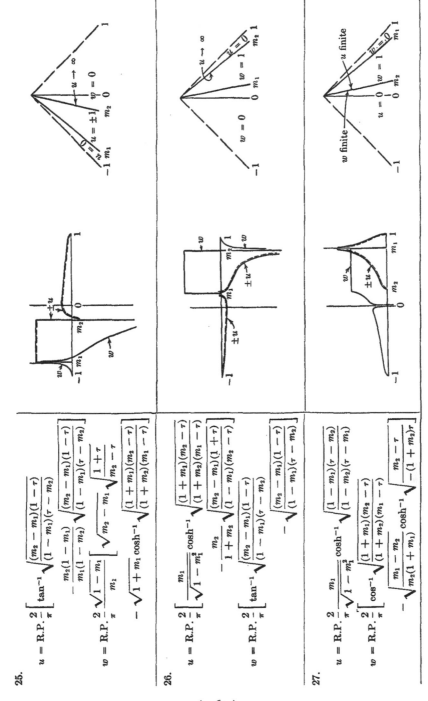

25.

$$u = \text{R.P.} \frac{2}{\pi}\left[\tan^{-1}\sqrt{\frac{(m_2-m_1)(1-\tau)}{(1-m_1)(\tau-m_2)}} - \frac{m_2(1-m_1)}{m_1(1-m_2)}\sqrt{\frac{(m_2-m_1)(1-\tau)}{(1-m_1)(\tau-m_2)}}\right]$$

$$w = \text{R.P.} \frac{2}{\pi}\left[\frac{\sqrt{1-m_1}}{m_1}\sqrt{m_2-m_1}\sqrt{\frac{1+\tau}{m_2-\tau}} - \sqrt{1+m_1}\cosh^{-1}\sqrt{\frac{(1+m_1)(m_2-\tau)}{(1+m_2)(m_1-\tau)}}\right]$$

26.

$$u = \text{R.P.} \frac{2}{\pi}\left[\frac{m_1}{\sqrt{1-m_1^2}}\cosh^{-1}\sqrt{\frac{(1+m_1)(m_2-\tau)}{(1+m_2)(m_1-\tau)}} - \sqrt{\frac{(m_2-m_1)(1-\tau)}{(1-m_1)(\tau-m_2)}}\right]$$

$$w = \text{R.P.} \frac{2}{\pi}\left[\tan^{-1}\sqrt{\frac{(m_2-m_1)(1-\tau)}{(1-m_1)(\tau-m_2)}} - \frac{m_2}{1+m_2}\sqrt{\frac{(m_2-m_1)(1+\tau)}{(1-m_1)(m_2-\tau)}}\right]$$

27.

$$u = \text{R.P.} \frac{2}{\pi}\left[\frac{m_1}{\sqrt{1-m_1^2}}\cos^{-1}\sqrt{\frac{(1+m_1)(m_2-\tau)}{(1+m_2)(m_1-\tau)}} - \sqrt{\frac{(1-m_1)(\tau-m_2)}{(1-m_2)(\tau-m_1)}}\right]$$

$$w = \text{R.P.} \frac{2}{\pi}\left[\frac{m_1-m_2}{m_2(1+m_1)}\cos^{-1}\sqrt{\frac{m_2-\tau}{-(1+m_2)\tau}}\right]$$

factor, the supersonic form may be written

$$
\left.
\begin{aligned}
x' &= x - \mu y \\
y' &= y - \mu x \\
z' &= \sqrt{1 - \mu^2}\, z
\end{aligned}
\right\}
\tag{13-5}
$$

where μ is the tangent of the angle between the axes (Fig. A,13c). This transformation is a special application of the well-known Lorentz transformation, and has the following properties:

1. If $f(x, y, z)$ is a solution of the steady flow equation (Eq. 10-4 or its equivalent expression in terms of one of the velocity components), then $f(x', y', z')$ is also a solution.
2. The Mach cone $x^2 = y^2 + z^2$ remains invariant under the transformation.
3. The x' and y' axes are oblique to each other and only the Mach angle is preserved.

In the plane of the wing, the conical variable τ becomes

$$
\tau(x', y', 0) = \frac{t - \mu}{1 - \mu}
\tag{13-6}
$$

where $t = \tau(x, y, 0)$.

If the variable τ is replaced by $\tau(x', y', z')$ in any expression for u or w in Table A,13a, it follows from (1) above that a new solution will have been generated. However, the transformation will hold only for the velocity component to which it was applied, since, from property (3), the relations (13-3 and 13-4) between the velocity components are not invariant under the transformation. Thus if the boundary conditions in u are to be satisfied in regions skewed according to Eq. 13-5, the new expression for w must be derived by the reapplication of Eq. 13-3, and vice versa if the boundary conditions are specified for w. For a more complete discussion, see [*142*].

METHOD OF APPLICATION OF CONICAL FLOW THEORY. The simplest use of the conical flow method is in finding the flow over wings of essentially conical form. Thus, following a preliminary application of the Prandtl-Glauert transformation (10-3), the pressure over a thin flat triangular wing at an angle of attack is given directly by solution 6 if the wing has subsonic leading edges, or by solution 3 if the leading edge is supersonic, when τ is replaced by y/x, its value in the $z = 0$ plane. Almost as direct is the solution for the flat rectangular wing of high aspect ratio (Fig. A,13d), in which each tip may be considered independently as long as the Mach cones from the extremities of the leading edge do not overlap.

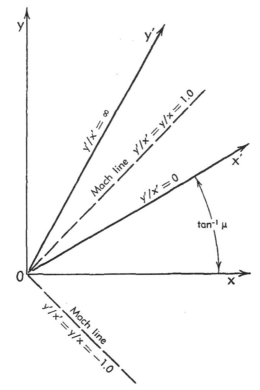

Fig. A,13c.   Oblique coordinates.

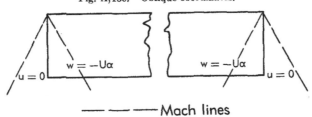

—————— Mach lines

Fig. A,13d.   Rectangular wing as composite of two independent
conical fields (solution 17, Table A,13a, with $m \to \infty$).

Between the tip regions, of course, the pressure is given by the two–
dimensional (Ackeret) theory.

In either of the cases cited, the fact that the entire wing lies ahead of
the Mach lines from points on the trailing edge makes it unnecessary to
consider the effect of terminating the infinite conical solution along that
edge, if one is interested only in the lift distribution. If the flow behind the
wing is also to be investigated, it becomes necessary to take into account

the fact that the wing is finite. By the method of superposition of conical flows the entire flow field behind such finite wings of simple planform can be found [154]. The procedure is the same as that by which the lift on finite wings of more general form may be derived from the infinite conical or two-dimensional solutions. A description of this procedure follows.[15]

Cancellation-of-lift procedure for finite wings. The basic consideration in determining the flow around the finite wing is that (except across a shock) there can exist no discontinuities in the pressure unsupported by a surface. Then the problem becomes the cancellation of such discontinuities introduced by the infinite conical flow field beyond the planform of the finite wing. This must be done without disturbing the boundary conditions elsewhere. In the case of a flat wing of specified camber, the latter provision means that the canceling flow must (i) induce no downwash within the boundaries of the finite wing and (ii) introduce no new lifting

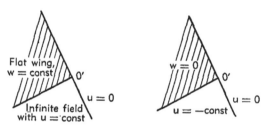

Problem field   Canceling field

Fig. A,13e.   Boundary conditions for cancellation
of uniform lift to derive a finite wing.

pressure outside these boundaries. It is generally necessary to consider also the character of the edge at which cancellation occurs—whether it is leading or trailing, subsonic or supersonic—since the solution is otherwise not unique (cf. solutions 18 and 19 of Table A,13a).

In the simplest problem, the lift to be canceled will be uniformly distributed over a semi-infinite region bounded by two straight lines, one the edge of the finite wing and one a ray of the infinite field (see Fig. A,13e). With conditions (i) and (ii) listed above, the boundary conditions of the cancellation field are then conical with respect to the intersection 0′ of the two lines. Thus the flow field due to angle of attack of the wing shown in Fig. A,13f may be derived from that of a triangular wing (after application of the Prandtl-Glauert transformation) by the superposition, with its origin at 0′, of the field given by solution 21, and at 0″ of the mir-

---

[15] The procedure presented here is superficially different from that first suggested in [142], and employed in [154] and subsequent papers using conical flow theory. However, the reader will see that either reduces to the other through an integration by parts.

ror image of such a field. These fields are characterized by a region of zero vertical velocity adjacent to a region of constant streamwise velocity and a finite value of u along the ray between. Since from solution 3 the pressure difference to be canceled is $2\rho_\infty U$ times

$$u_1 = \frac{mU\alpha}{\sqrt{m^2 - 1}} \qquad (13\text{-}7)$$

where α is the angle of attack of the wing and U the free stream velocity, the decrement in lifting pressure induced on the wing by the cancellation

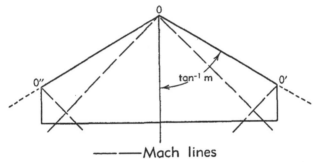

—— —— Mach lines

Fig. A,13f.   Wing with sweptback leading edge, straight trailing edge, and streamwise tips, at a high supersonic Mach number. (Plane transformed by Prandtl-Glauert transformation to $M_\infty = \sqrt{2}$; $m \gg 1$.)

process is, within the Mach cone of the right tip, $2\rho_\infty U$ times

$$\Delta u(x, y) = -\frac{2mU\alpha}{\pi \sqrt{m^2 - 1}} \tan^{-1} \sqrt{\frac{m(1 + t')}{-(1 + m)t'}}$$

where t' is the conical coordinate of $(x, y, 0)$ referred to $0'$ as origin. With a similar expression for the correction to the velocity in the left-hand tip cone, the problem of the lift on the depicted wing is completely solved.

However, the pressure to be canceled is not generally a constant, but varies, for example, in the case of a sweptback wing, with $t = \tau(x, y, 0)$ according to the streamwise component u in solution 3 or 6, depending on whether the leading edge is supersonic or subsonic. We therefore require an infinitesimal region of uniform lift (or u) with zero lift on either side, the lifting strip being bounded by two adjacent rays of slope t and $t + dt$ of the original conical field (Fig. A,13g) and an infinitesimal segment of the tip or trailing edge of the wing. Such a field may be obtained as the differential of a conical field with the appropriate boundary conditions. For the tip problem shown in Fig. A,13g, solution 21 is again employed, with the parameter m replaced by t, and the variable τ referred to the intersection $0'$ of the ray t with the tip as origin. Differentiating the resulting expression for u with respect to t would give the lifting pressure on the wing due to an infinitesimal strip with unit lift. The total decrement in lift at any point x, y within the Mach cone from the right tip would be $2\rho_\infty U$ times

$$\Delta u(x,\,y) = \frac{2\rho_\infty U}{\pi} \int_{t'=-1}^{t=m} u_\Delta(t) \left[\frac{\partial}{\partial t} + \frac{\partial}{\partial t'}\frac{dt'}{dt}\right] \tan^{-1} \sqrt{\frac{t(1+t')}{-(1+t)t'}}\; dt \quad (13\text{-}8)$$

in which $t' = \tau'(x,\,y,\,0) = (y - s)/[x - (s/t)]$ (see Fig. A,13g), u_Δ is given by solution 3 or 6, and the lower limit of integration corresponds to the value of t such that the Mach cone from $0'$ passes through the point x, y.

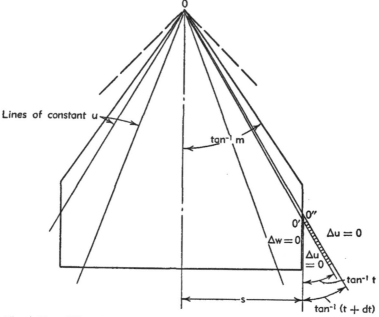

Fig. A,13g.   Wing of Fig. A,13f at a lower Mach number. (Plane transformed by Prandtl-Glauert transformation to $M_\infty = \sqrt{2}$; $m < 1$.)

If the leading edge of the wing is subsonic ($m < 1$), Eq. 13-8 is readily integrated.[16] If the leading edge is supersonic, the integral is not as tractable, and the source-sink methods to be described later appear more suited to the problem. In either case, the regions influenced by the two tips may overlap. Where this occurs, the effects of the cancellation of lift are of course additive.

To find the flow behind a straight or swept trailing edge, a similar procedure would be followed, employing solution 23 (see Fig. A,13h). If the angle of sweep is greater than the sweep of the Mach lines (Fig. A,13i), a portion of the wing will lie within the Mach cones of the cancel-

[16] The result is contained in a more general formula (13-37a) to be derived later.

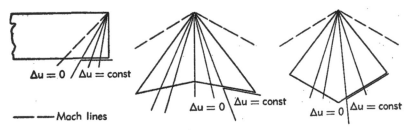

Fig. A,13h.   Wings with supersonic trailing edges, and the
conical flow fields required to cancel lift in the stream.

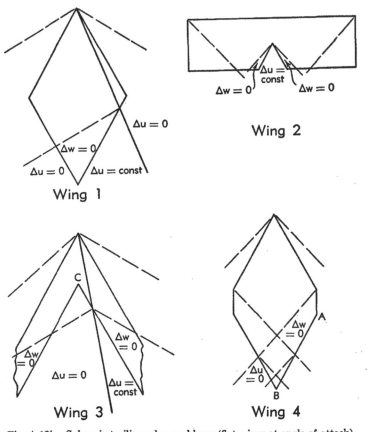

Fig. A,13i.   Subsonic trailing edge problems (flat wings at angle of attack).

ing fields, complicating the boundary conditions so that solution 23 will no longer apply. A uniform distribution of lift (wing 2) could be canceled behind a subsonically sweptback trailing edge without altering the slope of the streamlines ahead, by the application of solution 18 (or an oblique transformation of the same field) with opposite signs on upper and lower surfaces. The lift distribution on the wing would then be modified in such a way as to satisfy the Kutta condition. Cancellation of a nonuniform lift, however, would generally require displacement of the origin along the trailing edge, disrupting the conical form of the boundary conditions (wing 3). If the trailing edge is swept forward (wings 1 and 4) a similar problem arises or, in general, whenever another edge of the wing cuts

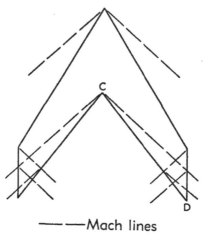

————Mach lines

Fig. A,13j.   Interference regions on a typical sweptback wing at $M_\infty = 1.5$.

across the Mach cone of a canceling field. In principle the resulting non-conical regions can be treated by the same superposition process as was the original problem, but in practice the resulting multiple integrals can seldom be evaluated analytically. If one edge lies entirely downstream of the other, as at A on wing 4, the lift distribution in the neighborhood of their intersection can be found in a single step by the lift cancellation method of [*141*] (discussed later in this article). When this is not the case, as at B on the same wing, and at C and D in Fig. A,13j, any cancellation of nonuniform lift at one edge will introduce errors at the other, so that such problems require an infinite number of steps for their complete solution by presently developed methods. Approximate solutions for the problem shown in Fig. A,13j and for the still more complex case (Fig. A,13k) in which both the tip and part of the leading edge lie within the region of influence of the trailing edge are discussed in [*155*] and later in this article.

Other applications.   The use of the conical flow solutions is not, of course, restricted to the cancellation of lift. In a later article, nonlifting components will be superimposed to derive finite wings of specified thickness distribution. The lifting solutions characterized by constant w over part of the x, y plane are useful in finding the effect of twist or of control surface deflection [*156,157*]. An interesting example of their application is

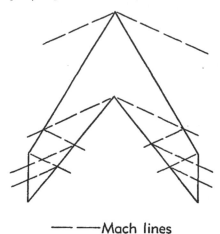

——— Mach lines

Fig. A,13k. Interference regions on the wing of Fig. A,13j at $M_\infty = 1.2$.

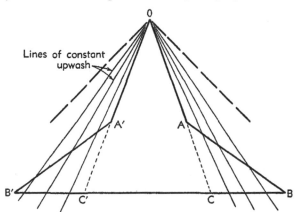

Fig. A,13l. Example of wing with "cranked" leading edge.

the wing with so-called "cranked" leading edge shown in Fig. A,13l, where the outer panels ABC and A'B'C' are operating in a conically varying upwash field. The solutions required [*158*] are of the same type as in the case of flap deflection.

Calculation of integrated forces and moments. If the planform, twist, and camber of a wing are derivable by the superposition of conical flows, the calculation of the total lift and pitching or rolling moment should also be found by consideration of the component fields. In each conical field, the pressure or u varies only with t, so that a single integration suffices to determine the lift or moment associated with that field. The resulting expressions may then be combined in exactly the same way as the velocity fields would be for the same wing. Since the result of the first integration is usually a simple algebraic expression, the conical flow method is often the most practical for the calculation of these integrated effects (see [*155*]).

SOURCE-DISTRIBUTION METHODS. BASIC FORMULAS. It is obvious that the conical flow method, though providing in certain cases very complete solutions in just one or two steps, must be restricted in its application to wings with rectilinear planforms and a minimum of interference between the edge flows. Somewhat more flexible tools are provided by the methods based on the solution (10-10) for a supersonic source, and related quantities.

For convenience of presentation, it will be assumed that the flow field has been normalized by the application of the Prandtl-Glauert transformation (10-3), so that the solution for the source takes the form

$$\varphi \sim \frac{1}{\sqrt{x^2 - y^2 - z^2}} \tag{13-9}$$

Puckett [*159*] has shown that the vertical velocity associated with this solution is zero everywhere in the $z = 0$ plane except at the location of the source itself, where the vertical velocity away from the plane is exactly π times the source strength per unit area. Thus the velocity potential φ at any point x, y, z, due to a distribution of sources in the horizontal plane, is

$$\varphi(x, y, z) = -\frac{1}{\pi} \iint \frac{w_u(\xi, \eta)d\xi d\eta}{\sqrt{(x - \xi)^2 - (y - \eta)^2 - z^2}} \tag{13-10}$$

The area of integration is the portion of the plane intercepted by the Mach forecone of the point x, y, z, and w_u is the vertical component of the velocity immediately above any source-point ξ, η, 0.

If, now, we consider the flow around a vertically symmetric wing—that is, an uncambered wing at zero angle of attack—the vertical velocity will be zero in the plane of symmetry everywhere except in the region occupied by the wing, where it will be determined by the shape of the wing surface. Thus the simple "thickness problem" is solved immediately by the integration of Eq. 13-10.

The expression for the streamwise component is obtained by the differentiation of Eq. 13-10 in the x direction and can be shown (VI,D,10) to reduce to

$$u(x, y, z) = \frac{1}{\pi} \int d\eta \int \frac{(x - \xi)w_u(\xi, \eta)}{[(x - \xi)^2 - (y - \eta)^2 - z^2]^{\frac{3}{2}}} d\xi \quad (13\text{-}11)$$

in which the bar on the inner integral sign signifies that the finite part is to be used (see VI,D,14.1). When the value of w_u outside of the wing is zero, the area of integration may be reduced to the part of the wing plan intercepted by the Mach forecone from $P(x, y, z)$ (see Fig. A,13m). Then integration of Eq. 13-11 by parts gives

$$u(x, y, z) = -\frac{1}{\pi} \left[\int_{AB} \frac{w_u(\xi, \eta)d\eta}{\sqrt{(x - \xi)^2 - (y - \eta)^2 - z^2}} \right.$$
$$\left. + \iint_{S} \frac{\partial w_u/\partial \xi}{\sqrt{(x - \xi)^2 - (y - \eta)^2 - z^2}} d\xi d\eta \right] \quad (13\text{-}12)$$

The first term of Eq. 13-12 is a line integral along the part of the leading edge intercepted by the Mach forecone from P. When the slope of the wing

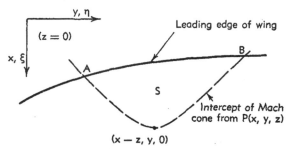

Fig. A,13m.   Region of integration for Eq. 13-12.

is constant in S, as in the case of a nonlifting wedge, the second term vanishes and only the line integral remains. For a surface consisting of several plane areas, superposition would result in as many line integrals along the ridge lines.

The expression of u in the form (13-12) seems first to have been noted by Evvard [160], and will be seen to have great practical value.

SOURCE DISTRIBUTION FORMULAS APPLIED TO THIN LIFTING WINGS WITH SUPERSONIC LEADING EDGES. If the boundary of the wing is supersonic within the area of integration, as for point x, y in Fig. A,13n, Eq. 13-10 may be evaluated directly and Eq. 13-12 will apply even in the case of a nonsymmetrical airfoil, because the airfoil can cause no disturbance in the field ahead of such a boundary, and w is consequently zero there. The upper and lower surfaces may be considered separately, since the only possibility of interaction is around a subsonic edge. Thus, outside the regions influenced by subsonic edges, we may write, for points on the upper surface,

$$\varphi(x, y, +0) = \frac{U}{\pi} \iint_S \frac{\alpha_u(\xi, \eta)d\xi d\eta}{\sqrt{(x - \xi)^2 - (y - \eta)^2}} \qquad (13\text{-}13)$$

and for the lower surface,

$$\varphi(x, y, -0) = -\frac{U}{\pi} \iint_S \frac{\alpha_l(\xi, \eta)d\xi d\eta}{\sqrt{(x - \xi)^2 - (y - \eta)^2}} \qquad (13\text{-}14)$$

where α_u and α_l are the local slopes of the upper and lower surfaces in the stream direction, and U is the free stream velocity. For a thin cam-

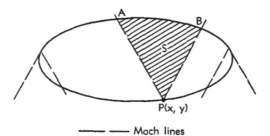

— — — Mach lines

Fig. A,13n. Example of the application of Eq. 13-12 to a finite lifting wing (arc AB entirely supersonic).

bered wing, $\alpha_u = \alpha_l$, and the difference in potential across the surface is simply

$$\Delta\varphi(x, y) = \frac{2U}{\pi} \iint_S \frac{\alpha(\xi, \eta)d\xi d\eta}{\sqrt{(x - \xi)^2 - (y - \eta)^2}} \qquad (13\text{-}15)$$

Evaluated for points along the trailing edges, Eq. 13-15 gives the span loading between the tip Mach cones in convenient form. The local lift, from Eq. 13-12, is proportional to

$$\Delta u(x, y) = -\frac{2U}{\pi} \left[\int_{AB} \frac{\alpha(\xi, \eta)d\eta}{\sqrt{(x - \xi)^2 - (y - \eta)^2}} \right.$$
$$\left. + \iint_S \frac{d\alpha/d\xi}{\sqrt{(x - \xi)^2 - (y - \eta)^2}} d\xi d\eta \right] \qquad (13\text{-}16)$$

Leading edge partially subsonic. If the boundary of the wing is subsonic anywhere within the Mach forecone of the point $x, y, \varphi(x, y)$ cannot in general be found by direct integration of Eq. 13-10. However, it may be (as in Fig. A,13l, with AB and $A'B'$ supersonic) that the upwash ahead of the wing is known, either from conical flow theory or otherwise. In such a case one may imagine the source sheet to cover the entire area of disturbance—that is, the entire region included by the Mach lines from O—and thus reduce the problem to the preceding case.

Using a variation of the above idea, Evvard [160] and Krasilshchikova [161] independently arrived at the general solution for a thin lifting wing of arbitrary planform, subject only to the condition that the foremost portion of the planform boundary be supersonic for some distance. Briefly, the development in this case is as follows:

Consider a point $P(x, y)$ near the right-hand tip of a wing of which the leading edge is largely supersonic. Let O (Fig. A,13o) be the point of the leading edge at which the leading edge becomes subsonic. Then the

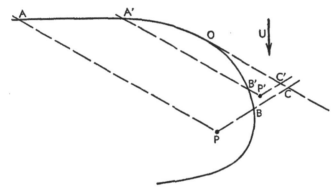

Fig. A,13o.　Region of integration for point near tip of wing with supersonic leading edge.

evaluation of the velocity potential at P by Eq. 13-10 includes integration over the area OCB of a function involving a vertical velocity generally undetermined in that region.

However, we may also write the expression for φ at a point P' of region OCB:

$$\varphi(x', y') = -\frac{1}{\pi} \iint\limits_{A'OC'P'} \frac{w(\xi, \eta)d\xi d\eta}{\sqrt{(x' - \xi)^2 - (y' - \eta)^2}} \qquad (13\text{-}17)$$

and this must equal zero if the point P' is ahead of the wing, or at any rate not in its wake. (When P' is in the wake, a different development is required.) If the integral equation obtained by setting $\varphi(x', y')$ equal to zero can be solved for $w(x', y')$, this quantity will be known both off and on the wing and Eq. 13-10 can be evaluated for point P.

The solution of the integral equation for w is made feasible by a transformation to a characteristic coordinate system, i.e. one with its axes in the direction of the Mach lines of the flow. The new variables r and s

(Fig. A,13p) are related to x and y by the equations[17]

$$r = (x - x_0) - (y - y_0)$$
$$s = (x - x_0) + (y - y_0)$$

(13-18a)

where x_0, y_0 are the coordinates of the point O. The coordinates of the source-point will again be designated by the Greek counterparts

$$\rho = (\xi - x_0) - (\eta - y_0)$$
$$\sigma = (\xi - x_0) + (\eta - y_0)$$

(13-18b)

With this transformation, the hyperbolic distance $\sqrt{(x - \xi)^2 - (y - \eta)^2}$ becomes $\sqrt{(r - \rho)(s - \sigma)}$, and the variables are then separable. The

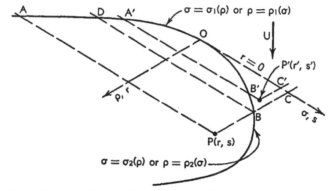

Fig. A,13p.   Characteristic coordinate system and equations of boundary curves.

element of area $d\xi\,d\eta$ transforms into $\frac{1}{2}d\rho\,d\sigma$. For point P' we obtain

$$\int_0^{r'} \int_{\sigma_1(\rho)}^{s'} \frac{w(\rho, \sigma)\,d\sigma\,d\rho}{\sqrt{r' - \rho}\,\sqrt{s' - \sigma}} = 0$$

(13-19)

where $\sigma = \sigma_1(\rho)$ is the equation of the supersonic portion of the leading edge.

Eq. 13-19 may be considered an Abel equation of the form

$$\int_0^{r'} \frac{F(\rho)\,d\rho}{\sqrt{r' - \rho}} = 0$$

(13-20)

for which the solution is $F(\rho) = 0$ provided that $F(\rho)$ is known to have no singularities in the region of integration. In the integrand of

$$F(\rho) = \int_{\sigma_1(\rho)}^{s'} \frac{w(\rho, \sigma)\,d\sigma}{\sqrt{s' - \sigma}}$$

(13-21)

[17] It may be well to recall at this time that the x, y coordinate system already includes a Prandtl-Glauert transformation, so that the Mach lines, and the new axes, are at 45° to the stream.

the only singularities are known to be of integrable order, so that $F(\rho)$ is itself nonsingular in the range $0 < \rho < r'$.

The integral equation $F(\rho) = 0$ can, in turn, be solved for w (which is *not* continuous, and is therefore not necessarily zero). The solution is of interest in itself as giving the upwash off the tip of a thin wing of arbitrary camber, but the explicit value is not needed for the calculation of the potential at P. Instead we write

$$F(\rho) = \int_{\sigma_1(\rho)}^{\sigma_2(\rho)} \frac{w(\rho, \sigma)d\sigma}{\sqrt{s' - \sigma}} + \int_{\sigma_2(\rho)}^{s'} \frac{w(\rho, \sigma)d\sigma}{\sqrt{s' - \sigma}} \tag{13-22}$$

$\sigma_2(\rho)$ being the subsonic portion of the wing boundary (to the right of O). Then

$$\int_{\sigma_2(\rho)}^{s'} \frac{w(\rho, \sigma)d\sigma}{\sqrt{s' - \sigma}} = -\int_{\sigma_1(\rho)}^{\sigma_2(\rho)} \frac{w(\rho, \sigma)d\sigma}{\sqrt{s' - \sigma}} \tag{13-23}$$

In the characteristic coordinates the velocity potential at P becomes (Fig. A,13p)

$$-\frac{1}{2\pi} \iint_{A\dot{O}\dot{C}P} \frac{w(\rho, \sigma)d\rho d\sigma}{\sqrt{(r - \rho)(s - \sigma)}} = -\frac{1}{2\pi}\left[\int_0^{\rho_2(s)} \frac{d\rho}{\sqrt{r - \rho}} \int_{\sigma_1(\rho)}^{\sigma_2(\rho)} \frac{w(\rho, \sigma)}{\sqrt{s - \sigma}}d\sigma\right.$$

$$+ \int_{\rho_2(s)}^{r} \frac{d\rho}{\sqrt{r - \rho}} \int_{\sigma_1(\rho)}^{s} \frac{w(\rho, \sigma)}{\sqrt{s - \sigma}}d\sigma$$

$$\left. + \int_0^{\rho_2(s)} \frac{d\rho}{\sqrt{r - \rho}} \int_{\sigma_2(\rho)}^{s} \frac{w(\rho, \sigma)}{\sqrt{s - \sigma}}d\sigma\right] \tag{13-24}$$

The first two terms in the expanded expression for $\varphi(x, y)$ are obtainable from the specification of the wing. The last term is obtainable from Eq. 13-23, with $s' = s$ (P' on BC, Fig. A,13p) and has the effect of canceling exactly the first term, so that

$$\varphi(x, y) = -\frac{1}{2\pi} \iint_{A\dot{D}\dot{B}P} \frac{w(\rho, \sigma)d\rho d\sigma}{\sqrt{r - \rho}\sqrt{s - \sigma}} \tag{13-25}$$

This result is identical with that obtained as Eq. 12-39 of VI,D, but is rederived here using the Evvard-Krasilshchikova approach as being more intuitive, and therefore more readily extended.

Eq. 13-25 leads to the interesting conclusion that the shape or existence of the surface ahead of the "reflected" Mach line BD has no effect on the flow at point P. Physically interpreted, it leads to the conjecture that the influence of a negative point disturbance on the under surface of a thin wing travels around the subsonic tip and somehow cancels the influence of the corresponding positive disturbance on the upper side within a region which may be thought of as resulting from the wrapping of the Mach cone from the under side around the side edge.

If the Mach forecone from P includes a second subsonic edge, as $O'A$ in Fig. A,13q, the upwash behind the Mach line from O' will cancel the influence of the region ahead of the "reflected" Mach line AE in the same way; and if the lines AE and BD intersect on the wing, the negative effects

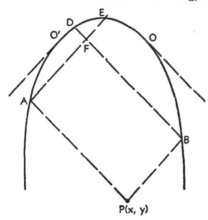

Fig. A,13q. Location of point and regions of integration for Eq. 13-26 and 13-27.

will be superimposed in the region ahead of the intersection. The net effect is to give, for a point located as in Fig. A,13q,

$$\varphi(x, y) = -\frac{1}{2\pi}\left[\iint\limits_{AFBP} \frac{w(\rho, \sigma)d\rho d\sigma}{\sqrt{r - \rho}\sqrt{s - \sigma}} - \iint\limits_{DEF} \frac{w(\rho, \sigma)d\rho d\sigma}{\sqrt{r - \rho}\sqrt{s - \sigma}}\right]$$

(13-26)

Eq. 13-26 may be differentiated to obtain the component of velocity proportional to the pressure. If $\sigma = \sigma_1(\rho)$ describes the portion of the planform boundary containing A, and $\rho = \rho_2(\sigma)$ the portion containing B, the expression for $\partial\varphi/\partial x$ may be written

$$u(x, y) = -\frac{1}{2\pi}\left\{\iint\limits_{AFBP} \frac{\partial w/\partial x}{\sqrt{r - \rho}\sqrt{s - \sigma}}d\rho d\sigma\right.$$

$$-\iint\limits_{DEF} \frac{\partial w/\partial x}{\sqrt{r - \rho}\sqrt{s - \sigma}}d\rho d\sigma + \int\limits_{DE} \frac{w(d\rho - d\sigma)}{\sqrt{r - \rho}\sqrt{s - \sigma}}$$

$$+\frac{1}{\sqrt{s - \sigma_1(r)}}\left[1 - \left(\frac{d\sigma_1}{d\rho}\right)_{\rho=r}\right]\int\limits_{\overrightarrow{EA}} \frac{w}{\sqrt{r - \rho}}d\rho$$

$$+\frac{1}{\sqrt{r - \rho_2(s)}}\left[1 - \left(\frac{d\rho_2}{d\sigma}\right)_{\sigma=s}\right]\int\limits_{\overrightarrow{DB}} \frac{w}{\sqrt{s - \sigma}}d\sigma\right\}$$ (13-27)

in which $\sigma_1(r)$ is the value of σ at the left-hand intersection A of the Mach forecone from P with the planform boundary and $\rho_2(s)$ is the value of ρ at the right-hand intersection (point B). The evaluation of the double integrals is facilitated by reference to Fig. A,13r, where the limits of integration are expressed in terms of the characteristic coordinates. If the location of P relative to the wing boundary is such that the lines AE and DB do not intersect on the wing, the second area integral is of course equal to zero, and the line integral along the arc $\overset{\frown}{DE}$ changes sign.

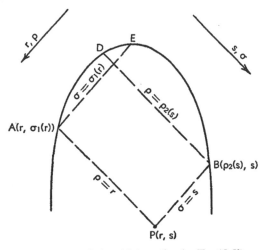

Fig. A,13r.   Limits of integration for Eq. 13-27.

If the point A falls on the supersonic portion of the leading edge, as in Fig. A,13p, the line AE vanishes, and Eq. 13-27 reduces to

$$u(x, y) = -\frac{1}{2\pi} \left\{ \iint_{ADBP} \frac{\partial w/\partial x}{\sqrt{r - \rho}\,\sqrt{s - \sigma}}\, d\rho\, d\sigma + \int_{\overset{\frown}{DA}} \frac{w(d\rho - d\sigma)}{\sqrt{r - \rho}\,\sqrt{s - \sigma}} \right.$$
$$\left. + \frac{1}{\sqrt{r - \rho_2(s)}} \left[1 - \left(\frac{d\rho_2}{d\sigma}\right)_{\sigma=s}\right] \int_{\overrightarrow{DB}} \frac{w\, d\sigma}{\sqrt{r - \rho}\,\sqrt{s - \sigma}} \right\} \quad (13\text{-}28)$$

If A or B is on a streamwise edge (tip), the corresponding one of the last two terms in Eq. 13-27 again drops out, since the coefficient becomes zero. In the case of a flat plate wing, the first two terms vanish and only line integrals remain to be calculated.

Pressure at point near subsonic trailing edge. If the Mach forecone from P includes a portion of the wake, as in Fig. A,13s, the foregoing development is no longer applicable. However, it may be applied to eliminate the region $OB'C$, where $\varphi = 0$, as before. It is then found that

⟨ 181 ⟩

the effect of the upwash in this region cancels the influence of any down-wash in region $D'OB'$, so that Eq. 13-27 may be applied as if the wing included the region of trailing vorticity behind it. The value of w in the region GBB' is now the only unknown quantity in the expression for $u(r, s)$. We may write Eq. 13-27 for point B, the intersection of the line $\sigma = s$ with the trailing edge, in terms of the downwash in the same region, and, according to the Kutta condition, set the resulting expression equal to zero. The equation thus obtained can then be written in the form of Eq. 13-20 and, following the procedure used in deriving Eq. 13-23, solved for

$$\int_{\sigma_t(\rho)}^{s} \frac{\partial w/\partial x}{\sqrt{s - \sigma}}\, d\sigma$$

in which $\sigma_t(\rho)$ is the value of σ along the trailing edge, in terms of ρ. This solution is all that is required to complete the expression of $u(r, s)$ in

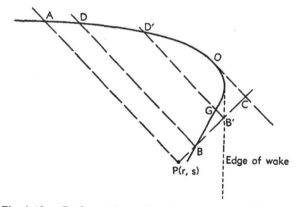

Fig. A,13s.   Regions of integration for point near trailing edge.

terms of known quantities, and has, moreover, the effect of canceling the contribution of the region $BDD'G$ to that expression, so that the final formula is simply (for P of Fig. A,13s)

$$u = -\frac{1}{2\pi}\left[\iint_{ADBP} \frac{\partial w/\partial x}{\sqrt{r - \rho}\sqrt{s - \sigma}}\, d\rho d\sigma + \int_{DA} \frac{w(d\rho - d\sigma)}{\sqrt{r - \rho}\sqrt{s - \sigma}}\right]$$

$$(13\text{-}29)$$

If the leading edge at A is subsonic, or if the line $\rho = r$ from P intersects the wing tip, as in Fig. A,13r, a similar derivation will yield Eq. 13-27 with the last term omitted. If the point A, as well as B, lies on a subsonic trailing edge, both the last and the next-to-last terms in Eq. 13-27 will drop out.

Eq. 13-29 is directly derivable from Eq. 13-53 of VI,D but it is hoped that the present derivation will lend itself more readily to extension by the reader to other cases. Evvard [160] has carried through an even more general derivation, in that the slopes of the upper and lower surfaces of the wing are allowed to vary independently, so that any distribution of thickness, as well as camber and twist, may be taken into account. In the present treatment, however, thickness effects will be considered separately (see Art. 14), in line with usual linear theory practice.

Table A,13b summarizes the formulas derived in the foregoing section and indicates the conditions under which they apply. Also indicated is a situation (region VII) for which further analysis is required. While the Evvard-Krasilshchikova method can be extended to cover such regions, it is probable that the full range of its usefulness per se has been explored. However, several of the basic ideas have been incorporated into a highly practical cancellation method first suggested by Goodman [162] and developed by Mirels [141]. This method is particularly useful for the treatment of the part of the wing influenced by a subsonic trailing edge. Therefore we will not consider further extensions of the preceding method, but will assume that the formulas given will be supplemented by the use of the Goodman-Mirels cancellation technique, a discussion of which follows.

CANCELLATION OF LIFT BY SOURCE DISTRIBUTIONS. The basic concept of the cancellation method has already been discussed, and in connection with the present problem of finding the lift distribution over a wing of given shape may be summarized as follows:

1. The lift distribution must be known over some wing of which the specified wing forms a part.
2. A flow must be formulated which (a) coincides in lift distribution with that calculated over the regions of the plane outside of the plan area of the specified wing, (b) introduces no vertical velocity in the area occupied by the specified wing, and (c) introduces no lift outside the boundaries of the large wing. Note that the original calculation may be made by any method—conical flows, Evvard-Krasilshchikova, or any other—so that the present method is no longer limited to wings with supersonic leading edges.

The general formula for a flow in which the downwash is specified over part of the plane and the lift over the remainder has been derived in VI,D. If the dividing line is a subsonic trailing edge of the wing of which the pressure distribution is being sought, so that the continuity of pressure across it is a boundary condition of the problem, the formula for the local lift is given explicitly. From VI,D (Eq. 12-40) we obtain, by setting w equal to zero on the specified wing,

$$\Delta u(r, s) = - \frac{\sqrt{r - \rho_2(s)}}{\pi} \int_{\rho_1(s)}^{\rho_2(s)} \frac{u(\rho, s)d\rho}{(r - \rho) \sqrt{\rho_2(s) - \rho}} \qquad (13\text{-}30)$$

Table A,13b. Formulas for lifting pressure on a wing with the foremost portion of the leading edge supersonic. (Formulas derived by the Evvard-Krasilshchikova method.)

Points in region	Mach forecone intersects[1]	Formula for u
I	supersonic edges only[2]	13-16
II	1 subsonic leading edge, 1 supersonic leading edge	13-28
III	2 subsonic leading edges[3]	13-27
IV	1 subsonic trailing edge,[3] 1 supersonic leading edge	13-29
V	1 subsonic trailing edge,[3] 1 subsonic leading edge	13-27, without last term
VI	2 subsonic trailing edges[3]	13-27, without last two terms
VII	2 interacting subsonic edges	Not presented. See discussion on p. 188.

[1] Number of intersections limited to two.
[2] Leading edge assumed supersonic throughout region I.
[3] or streamwise edge.

where $\Delta u(r, s)$ is the incremental streamwise velocity, proportional to the loss in lift, induced at the point r, s on the specified wing by the cancellation flow; $\rho_1(s)$ (see Fig. A,13t) is the value of ρ at the intersection A of the line $\sigma = s$ with the leading edge of the wing for which the lift is known; and $\rho_2(s)$ is the value of ρ at the intersection B of $\sigma = s$ with the trailing edge of the specified wing. (The lift to be canceled is assumed to occur to the right of the specified wing. If not, r and s are simply interchanged in Eq. 13-30 and ρ replaced by σ.)

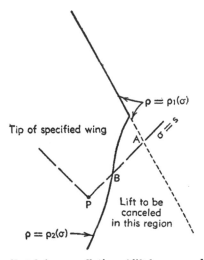

Fig. A,13t.   Sketch for cancellation of lift by source distribution.

If the boundary $\rho = \rho_2(\sigma)$ of the wing beyond which the lift is to be canceled is a leading edge, the formula for the pressure is considerably more complicated. For this case we must differentiate

$$\Delta\varphi = -\frac{\sqrt{r - \rho_2(s)}}{\pi} \int_{\rho_1(s)}^{\rho_2(s)} \frac{\varphi d\rho}{(r - \rho)\sqrt{\rho_2(s) - \rho}} \qquad (13\text{-}31)$$

which is Eq. 12-41 of VI,D with the condition of zero vertical velocity on the specified wing inserted. The result (see Appendix of [141]) may be written in the form

$$\Delta u = -\frac{1}{2\pi}\left[\sqrt{r - \rho_2(s)}\int_{\rho_1(s)}^{\rho_2(s)} \frac{2u(\rho, s)d\rho}{(r - \rho)\sqrt{\rho_2(s) - \rho}}\right.$$
$$\left. - \frac{1 - (d\rho_2/d\sigma)_{\sigma=s}}{\sqrt{r - \rho_2(s)}} \int_{\rho_1(s)}^{\rho_2(s)} \frac{(u - v)d\rho}{\sqrt{\rho_2(s) - \rho}}\right] \qquad (13\text{-}32)$$

in which v is the sidewash associated with the original wing flow, obtainable from the streamwise component of velocity u by the relation

$$v = \frac{\partial}{\partial y} \int_{x_1(y)}^{x} u(x, y)dx \tag{13-33}$$

where $x = x_1(y)$ is the leading edge of the original wing. The second term of Eq. 13-32 introduces the expected suction peak at the leading edge into the pressure distribution, and is important in computing the drag (Art. 2).

As an illustration of the use of the Goodman-Mirels method, yielding a result also of considerable interest in itself, consider the derivation of the formula for the lifting pressure at a point near the arbitrarily curved

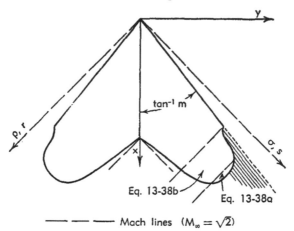

Fig. A,13u. Wing with straight swept leading edge and curved tips, for which lift distribution is given by Eq. 13-34 and 13-38.

tip of a flat wing with sweptback leading edge (Fig. A,13u), subject only to the following restrictions:

1. At the root, the leading edge of the wing consists of two straight line segments lying within the Mach cone from their juncture.
2. The remainder of the leading edge and the tips must also be subsonic at all points and lie within the infinite triangle formed by the rearward extension of the straight portions of the leading edge.
3. The Mach forecone from the point at which the pressure is to be calculated intersects only, on the left, the straight portion of the leading edge and, on the right, the nearby tip (or curved portion of the leading edge). If the point of intersection with the tip is one at which the wing is entering the stream, there is the further restriction that it is not directly downstream of some other part of the wing (as in a re-entrant side edge).

The basic solution is that for the infinite triangular wing at an angle of attack. From solution 6, Table A,18a, we obtain for the value of the u component of the velocity in the plane of the wing (after applying the Prandtl-Glauert transformation),

$$u_\Delta = \frac{mu_0x}{\sqrt{m^2x^2 - y^2}} \qquad (13\text{-}34)$$

in which m is the tangent of the semiapex angle,

$$u_0 = \frac{mU\alpha}{E'(m)} \qquad (13\text{-}35)$$

is the value of u_Δ along the axis of symmetry, $E'(m)$ is the elliptic integral, U is the stream velocity, and α the angle of attack. The velocity u_Δ is to be canceled in the region shown shaded in Fig. A,13u.

The expression for u_Δ, rewritten in terms of ρ and σ, with the origin at the apex of the wing, is

$$u_\Delta = \frac{mu_0(\sigma + \rho)}{\sqrt{m^2(\sigma + \rho)^2 - (\sigma - \rho)^2}} \qquad (13\text{-}36)$$

and the right-hand leading edge of the triangular wing is given by

$$\rho = \rho_1(\sigma) = \frac{1 - m}{1 + m}\sigma \qquad (13\text{-}37)$$

Substituting these expressions into Eq. 13-30 and integrating gives [152]

$$\Delta u(r, s) = -u_\Delta \Lambda_0(\psi, k) + \frac{u_0}{\pi}\sqrt{\frac{m[r - \rho_2(s)]}{s}}\, K(k) \qquad (13\text{-}38\text{a})$$

$$k = \frac{1}{2}\sqrt{\frac{1 - m}{ms}[(1 + m)\rho_2(s) - (1 - m)s]}$$

$$\psi = \tan^{-1}\sqrt{\frac{(1 + m)s - (1 - m)r}{(1 - m)[r - \rho_2(s)]}}$$

$$\Lambda_0(\psi, k)^{18} = \frac{2}{\pi}\{K(k)E(\psi, k') - [K(k) - E(k)]F(\psi, k')\}$$

for a point of which the Mach forecone cuts a trailing edge; that is, the slope $\rho_2'(s)$ exceeds or is equal to one. If $\rho_2'(s) < 1$, Eq. 13-32 applies, and the following *additional* term results:

$$\frac{u_0}{\pi}[1 - \rho_2'(s)][2E(k) - K(k)]\sqrt{\frac{s}{m[r - \rho_2(s)]}} \qquad (13\text{-}38\text{b})$$

When the point $\rho_2(s)$, s lies within the vortex sheet originating at some portion of the wing ahead, the sidewash v in Eq. 13-32 is affected so that Eq. 13-38 no longer holds. However, the problem may still be solved with

[18] Tabulated in [161,163].

the aid of Eq. 13-37. A simple example of such a case is worked out in [141].

The Goodman-Mirels method is particularly useful in finding the pressure near the trailing edge when the variation of lift ahead is already somewhat complex. For example, let us consider region VII of the wing shown in Table A,13b. The new element in region VII is the effect of the reflection, at the trailing edge ahead of the region, of the Mach wave bounding the region of influence of the far tip. This influence is expressed by the difference between the velocity in region V and that in region IV (or equally between $u(x,y)$ in region III and $u(x,y)$ in region II), and is to be canceled in the region between the trailing edge and the extension of the Mach line separating those regions. Thus the expressions for u, ρ_1, and ρ_2 can be determined for substitution into Eq. 13-30 and the increment u found for extending the solution from region V into region VII.

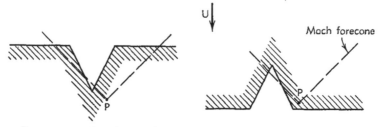

Re-entrant leading edge   Re-entrant trailing edge

Fig. A,13v. Problems not readily solved by lift-cancellation methods.

As indicated in the discussion of the conical flow method, the foregoing procedure may be applied any number of times on alternate sides of a wing, so that in principle the pressure on any wing, of no matter how low aspect ratio, may be completely determined, providing only that the trailing edge eventually has a supersonic segment. Where the wing terminates in an angle formed by two subsonic edges (Fig. A,13i) the process has no end, but may be carried to a point where the region of undetermined pressures is arbitrarily small. A solution in closed form would require the inversion of an integral equation involving more than the two regions considered in the derivation of Eq. 13-30, and this has not as yet been attempted for the general case (however, see [148,164]).

If two subsonic leading or trailing edges meet in a re-entrant angle, the Mach forecones of points on the wing are again divided into a number of disconnected or irregular regions (see Fig. A,13v). However, whereas in the preceding problem the errors introduced by ignoring the more remote regions took the form of extraneous lift, which could be eliminated by successive applications of the same lift-cancellation procedure, in the problems of Fig. A,13v the errors would involve unwanted variations of downwash on the wing, so that a somewhat different procedure must be followed.

For the problem of re-entrant trailing edge, the lift cancellation procedure may be carried out in a reverse direction, so to speak, to obtain an expression for the pressure at any point in the region of the trailing edge. Again an infinite number of terms is required for a complete solution, but an arbitrary degree of accuracy may be obtained with a finite number. Thus (Fig. A,13w) the lift at a point P on one side of the wing is expressed by Eq. 13-30 in terms of the known value of u in the region to be cut out and of the as-yet-undetermined value of u at points P_1 on the other side of the cutout. The value of u at P_1 is, in turn, expressed in terms of u at

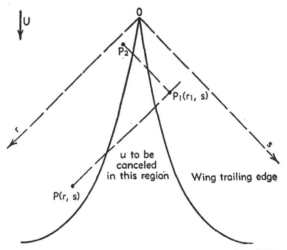

Fig. A,13w. Successive terms in pressure at point within Mach cone from apex of sweptback trailing edge.

the points P_2 on the first side of the wing, and so on. At any step in this procedure it may be assumed that the effect of the as-yet-undetermined pressures will be negligible and the series terminated. This procedure has the advantages that even in the first approximation, obtained by assuming Δu to be zero on the opposite wing panel, the Kutta condition will be satisfied at the trailing edge, and that the only violation of the boundary conditions occurs at points relatively remote from the point at which the pressure is being determined. If the trailing edge cutout occupies only a small portion of the Mach cone from its apex, a second or third step may be required for satisfactory accuracy.

It is apparent that, in the important case of a flat sweptback wing with straight leading edges, applying the foregoing procedure to Eq. 13-34 gives, even in the first approximation, incomplete elliptic integrals of the third kind, functions which have not as yet been tabulated. For engineering purposes, further simplification will generally be desirable. Actually, the lift to be canceled between the branches of the trailing edge differs very little from $2\rho_\infty U u_0$, the value along the axis of symmetry. In the following approximation, the Kutta condition is satisfied by taking into account the actual value of the lift to be canceled at the trailing edge near the point at which the pressure is to be calculated. Furthermore, the variation of lift is closely approximated along the forward Mach line ($\rho = r$, Fig. A,13x) up to a point r, s_0 at which the approximate value equals $2\rho_\infty U u_0$. The latter value is then assumed for the remainder of the

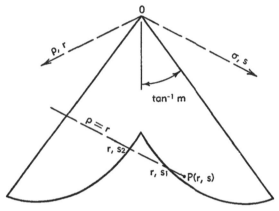

Fig. A,13x. Sketch for Eq. 13-39.

wake region. This procedure results in the following expression for the trailing edge effect on u:

$$\Delta u(r, s) = -\frac{2u_0}{\pi}\left[\tan^{-1}\sqrt{\frac{s_1 - s_2}{s - s_1}} - \tan^{-1}\sqrt{\frac{s_1 - s_0}{s - s_1}}\right.$$
$$\left. + \sqrt{\frac{(1 + m)r - (1 - m)s_0}{(1 + m)r - (1 - m)s}}\,\tan^{-1}\sqrt{\frac{(1 + m)r - (1 - m)s}{(1 + m)r - (1 - m)s_0}\left(\frac{s_1 - s_0}{s - s_1}\right)}\right]$$

$$(13\text{-}39)$$

Eq. 13-39 is written for a point on the right-hand wing, as shown in Fig A,13x. The origin of the coordinate system is at the leading edge apex. The quantities s_1 and s_2 are the σ coordinates (see Eq. 13-18) of the near and far intersections of the line $\rho = r$ with the trailing edge of the wing, and s_0 has the value

$$s_0 = \frac{(1 - m^2)rs_1 - (r - ms_1)^2}{(1 - m)[(1 + m)s_1 - (1 - m)r]}$$

Eq. 13-39 applies to an arbitrarily curved trailing edge. If the trailing edge is composed of straight lines, the first term in the equation, which approximates the effect of canceling the uniform lift $2\rho_\infty U u_0$ throughout the wake region, may be replaced by the exact value given by solution 18 of Table A,13a. Thus, for a simple V-shaped wing with subsonic leading and trailing edges, the reduction in u due to the trailing edge is

$$\Delta u(r, s) = - u_0 \left\{ \frac{F(\phi, \sqrt{1 - m_t^2})}{K(\sqrt{1 - m_t^2})} - \frac{2}{\pi} \left[\tan^{-1} \sqrt{\frac{s_1 - s_0}{s - s_1}} \right. \right.$$

$$\left. \left. - \sqrt{\frac{(1 + m)r - (1 - m)s_0}{(1 + m)r - (1 - m)s}} \tan^{-1} \sqrt{\frac{(1 + m)r - (1 - m)s}{(1 + m)r - (1 - m)s_0} \left(\frac{s_1 - s_0}{s - s_1} \right)} \right] \right\}$$

$$\phi = \sin^{-1} \frac{1}{x - c_0} \sqrt{\frac{(r - c_0)(s - c_0)}{1 - m_t^2}} \tag{13-40}$$

where m_t is the inclination of the trailing edge to the stream and c_0 is the length of the root chord.

Eq. 13-40 gives accuracy varying from complete agreement at the trailing edge to a maximum error of 1.5 per cent when compared with the trailing edge corrections given in [155]. Eq. 13-39 gives results within 8 per cent of the correct values.

A,14. Specific Planforms. Lift Distribution, Lift and Drag Due to Angle of Attack. This article will present results of the application of the foregoing procedures to specific planforms. The planforms considered will be limited to those bounded by straight lines, although the trailing edges, if supersonic, may be of almost any shape. Only the effect of angle of attack (the flat plate problem) and unyawed wings will be considered. Calculated loadings will be shown in plots of $C_p = (p_{lower} - p_{upper})/\frac{1}{2}\rho_\infty U^2 = 4u/U$, and formulas for the lift, drag, and aerodynamic center location will be given. The formulas will be written in terms of the physical dimensions of the wing—root chord c_0, span b or semispan s, and aspect ratio $\mathcal{R}$—so that the Prandtl-Glauert transformation factor $1/\beta$ ($\beta = \sqrt{M_\infty^2 - 1}$) will now appear with all the streamwise quantities in the previously derived expressions. Then, since the pressure is proportional to the streamwise gradient of φ, the formulas for lift will contain an additional factor of $1/\beta$.

Rectangular wings. The solution for the high aspect ratio rectangular wing is most readily obtained by the superposition of two conical fields (solution 21, Table A,13a) on a cylindrical field (Ackeret's solution for a flat plate). The resulting form of the lift distribution is shown in spanwise sections in Fig. A,14a. Also shown are experimental values obtained [165] on the flat upper surface of a wing with half-diamond profile (see figure) in tests at a Mach number of 2.

If either the Mach number or aspect ratio is low enough to cause the tip Mach cones to overlap, the negative effects are additive in the region of overlapping (region III in sketch, Fig. A,14b). The tip effects add up to a constant all along the line $x = 2s$ joining the points of intersection of the Mach lines with the opposite tips, and combine to reduce the lift to zero at that streamwise station (a distance $1/\beta$ times the span downstream of the leading edge). Behind this line the lift is negative.

Reflection of the tip Mach waves at the opposite side edges has the

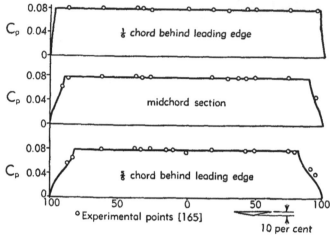

Fig. A,14a.   Distribution of lift on a rectangular flat plate ($\mathcal{R} = 5.7$) at 4° angle of attack and Mach number of 2.0.

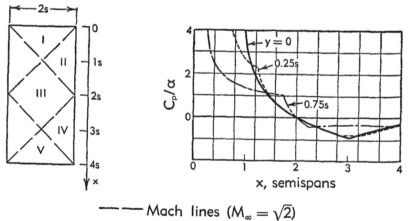

Fig. A,14b.   Chordwise lift distribution at various spanwise stations of rectangular wings [167].

effect of restoring some of the lift. The magnitude of this effect (according to the simplified theory being applied) may be calculated about equally well by conical-flow [168] or source-distribution [167] methods. It should be noted, however, that early separation of the flow and the formation of vortices around the side edges (Art. 1) make the flow near the tips rather different from that assumed, so that the results of this simplified theory are of limited interest.

The loading on rectangular wings has been calculated as far downstream as the intersections of the reflected Mach waves with the side edges. The results are presented for three chordwise sections in Fig. A,14b. Discontinuities in slope mark the location of the waves and reflected waves from the tips. It is seen that at the farthest station downstream the lift is still negative.

The total lift on the rectangular wing with aspect ratio $R \geq 1/\beta$ is readily found by subtracting the integrated tip losses from the lift for the infinite wing. In coefficient form we get

$$C_{L_\alpha} = \frac{4}{\beta}\left(1 - \frac{1}{2\beta R}\right), \qquad \text{for } \beta R \geq 1 \qquad (14\text{-}1a)$$

For $\beta R < 1$, application of the conical flow method gives [168]

$$C_{L_\alpha} = \frac{4}{\pi\beta}\left[\left(2 - \frac{1}{\beta R}\right)\sin^{-1}\beta R + (\beta R - 2)\cosh^{-1}\frac{1}{\beta R}\right.$$
$$\left. + \left(1 + \frac{1}{\beta R}\right)\sqrt{1 - \beta^2 R^2}\right] \qquad (14\text{-}1b)$$

$$\tfrac{1}{2} \leq \beta R \leq 1$$

The variation of lift curve slope with Mach number for rectangular wings of various aspect ratios is shown in Fig. A,14c. In Fig. A,14d, the variation of aerodynamic center location, $x_{a.c.}$, with Mach number is shown, obtained from the same source [168]. The drag due to angle of attack, as of any flat wing with supersonic leading edge, is merely equal to the lift times the angle of attack.

Trapezoidal wings. The high aspect ratio trapezoidal wing is treated very similarly to the rectangular. On the two-dimensional field we superimpose, in the case of subsonically raked-in tips, a field obtained from solution 22, Table A,13a, by letting $m_1 \rightarrow -\infty$. If the tips are raked out, the conditions for a subsonic edge will be satisfied by solution 8; a solution for supersonic edges can be derived from solution 1 by subtracting the given field from one of the same form, with infinite m. With supersonically raked-in tips, the pressure has the two-dimensional value all over the wing.

If the side edges are subsonic, reflection of the tip Mach waves will again occur at low Mach numbers, on low aspect ratio wings. If the edges are raked in, or trailing, Eq. 13-30 provides a means of calculating the incremental lift. For raked-out edges, the solution has been obtained by a source-distribution method in the previously cited work [167] by Behrbohm.

Section lift distributions for this last case are reproduced in Fig.

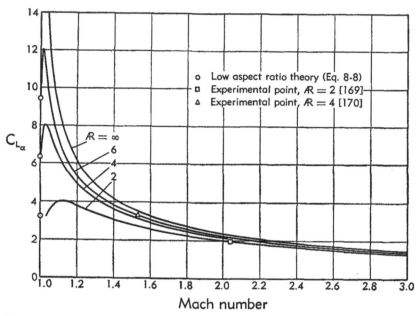

Fig. A,14c. Theoretical variation with Mach number of the
lift curve slope of rectangular wings.

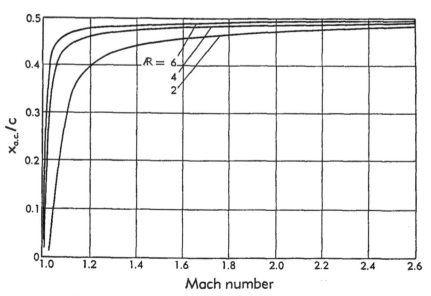

Fig. A,14d. Theoretical variation with Mach number of the
aerodynamic center location of rectangular wings.

A,14e. In addition to showing the expected suction peaks at the raked-out side edges, the loading differs from that on the rectangular wing by remaining positive at all points. Discontinuities in pressure gradients again indicate the influence of the tips. If the side edges were extended beyond the tip Mach lines, the suction peaks would be replaced by plateaus in the loading, covering the regions between the tip Mach lines and the side edges.

As a result of the reversibility property of the total lift (Art. 6), only two cases need be considered in determining the lift curve slopes for

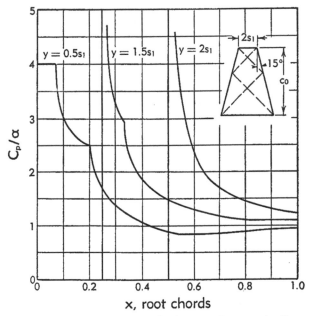

Fig. A,14e. Section lift distributions on a low aspect ratio
trapezoidal wing with raked-out subsonic tips [167].

trapezoidal wings: (1) subsonically raked and (2) supersonically raked tips. In both cases, the raked-in tip presents the simpler approach. If the side edge lies ahead of the tip Mach cone, the lift is uniform over the entire wing and the lift coefficient is identical with that for infinite aspect ratio, so that

$$C_{L_\alpha} = \frac{4}{\beta}$$

If the side edge is subsonic, subtracting the tip-induced losses in lifting pressure from the two-dimensional lift gives

$$C_{L_\alpha} = \frac{4}{\beta}\left[1 - \frac{(1 - \beta\tan\delta)c_0}{2\beta\bar{b}}\right], \qquad \bar{b} \ge \frac{c_0}{\beta} \qquad (14\text{-}2)$$

in which δ is the angle of inclination of the side edge to the stream and $\bar{b}$ is the width of the wing at midchord. For aspect ratios and Mach numbers such that $\beta\bar{b} < c_0$, no closed expression for C_{L_α} has been derived. Numerical values for two values of $\beta \tan \delta$ have been calculated by Behrbohm [167] and are shown in Fig. A,14f, together with the curve for $\beta \tan \delta = 0$, the rectangular wing.

In Fig. A,14g, aerodynamic-center locations are presented for the same families of wings, flying with the longer base trailing. Numerical values for $\beta\bar{b} < c_0$ are again taken from [167]. If $\beta\bar{b} \geqq c_0$, the aerodynamic center is readily located by resolving the loading into its component

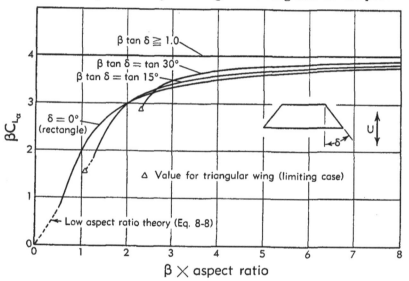

Fig. A,14f.   Lift curve slope of trapezoidal wings.

parts: the uniform two-dimensional loading and the conical tip corrections, the latter having their centroids at the centroids of area of the tip Mach cones. Results for both raked-in and raked-out trapezoids in such cases are included in the figure.

The drag due to lift of the trapezoidal wing with raked-in tips is simply equal to the lift times the angle of attack, since there are no subsonic leading edges. The same is true if the tips are raked out through more than the Mach angle. However, if the side edges are raked out but lie within the Mach cones from the leading edge, the drag will be reduced by a suction force calculated by means of Eq. 12-1 as follows:

From solution 8, Table A,13a, the streamwise component of velocity u approaches infinity at the tip as

$$\frac{2U\alpha m}{\pi(1 + m)\beta} \sqrt{\frac{x + \beta y}{mx - \beta y}} \qquad (14\text{-}3)$$

where m is β times the inclination of the side edge to the stream, and the origin of the coordinate system is at the tip of the leading edge. Then

$$C_x = \frac{2U\alpha\sqrt{mx}}{\pi\beta\sqrt{1+m}} \tag{14-4}$$

$$\frac{dT}{dx} = \frac{4\rho_\infty U^2\alpha^2 mx}{\pi\beta^2}\sqrt{\frac{1-m}{1+m}} \tag{14-5}$$

and the total thrust is

$$T = \frac{4\rho_\infty U^2\alpha^2 m}{\pi\beta^2}c_0^2\sqrt{\frac{1-m}{1+m}} \tag{14-6}$$

as long as the tip Mach cones do not intersect the opposite side edges. If this condition is violated ($c_0/\beta > \bar{b}$, the average span), the strength of

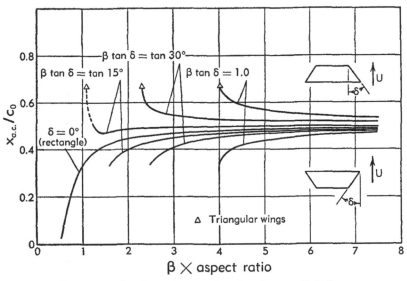

Fig. A,14g. Aerodynamic center location of trapezoidal wings.

the singularity along the edge will be modified by the interfering tip flow. The effect of the interference would not be difficult to calculate, but is not of great practical interest, being somewhat unrealistic. The variation of drag rise with aspect ratio and rake angle is shown in Fig. A,14h.

Pointed wings (triangular, diamond-shaped and swallowtail planforms). The loading on any wing with supersonic trailing edges and tips tapered to a point can be represented by a single conical expression depending only on the relative sweep of the leading edge and the Mach lines. The expression is, of course, that for the corresponding triangular wing, which is merely a special example of such wings. At low Mach numbers, such that the foremost Mach lines lie ahead of the leading edge, the loading is proportional to the incremental velocity given by Eq. 13-34 and shown as solution 6 in Table A,13a. In chordwise sections the lift distribution has the form shown in Fig. A,14i, where section lift distributions are plotted for a slender triangle. It is of interest to note that the span loading on the triangular wing is elliptic when the leading edge is subsonic.

Experiments on triangular wings at supersonic speeds have given pressure distributions in good agreement with the linear theory at very low angles of attack. At higher angles of attack, particularly when the aspect ratio is very low, the flow separates at the leading edge and the simple inviscid fluid theory no longer applies. The data shown in Fig. A,14i were obtained at $M_\infty = 1.37$ with a wing having a 40° apex angle

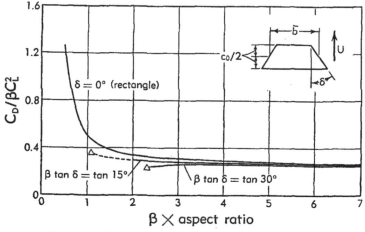

Fig. A,14h. Variation of drag rise factor with aspect ratio and rake angle for trapezoidal wings.

(70° sweepback of the leading edge) and 6 per cent thick, biconvex sections with a rounded leading edge. The effects of separation at the leading edge are evident even at a 3° angle of attack.

At higher Mach numbers, the flow normal to the leading edge becomes supersonic and the suction peaks observed in Fig. A,14i are replaced by the two-dimensional plateau in the loading (solution 3, Table A,13a). Some experimental lift distributions have been published [171] for triangular wings at such speeds, but the thickness of the wings (10 per cent) combined with the high Mach number, resulted in prominent second order effects.

General formulas for the total lift of pointed wings have been derived by Puckett and Stewart [172]. If m is β times the cotangent of the angle of sweepback of the leading edge and ξ the ratio of the cotangent of the angle of sweepback of the leading edge to that of the trailing edge, the lift curve slope is[19]

$$C_{L_\alpha} = \frac{4m}{\beta E'(m)} \left[\frac{\xi}{1+\xi} + \frac{1-\xi}{(1-\xi^2)^{\frac{1}{2}}} \cos^{-1}(-\xi) \right] \qquad (14\text{-}7)$$

when $m \leqq 1$, and

$$C_{L_\alpha} = \frac{8m}{\beta\pi(1+\xi)} \left[\frac{1}{\sqrt{m^2-\xi^2}} \cos^{-1}\left(-\frac{\xi}{m}\right) + \frac{\xi}{\sqrt{m^2-1}} \cos^{-1}\frac{1}{m} \right] \qquad (14\text{-}8)$$

when $m \geqq 1$. Diamond-shaped wings are included by permitting ξ to take on negative values.

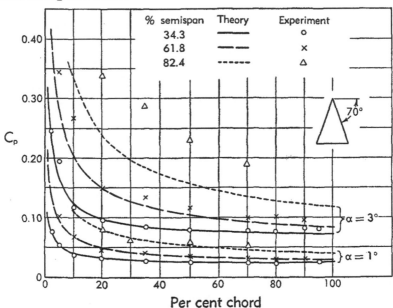

Fig. A,14i. Variation of lifting pressure along three streamwise sections of a triangular wing with 40° apex angle, at $M_\infty = 1.37$; theory and experiment [173].

Values of βC_{L_α} are plotted against m for various values of ξ in Fig. A,14j, and the corresponding center-of-pressure locations in Fig. A,14k. Positive values of ξ correspond to wings with cutouts at the relatively inefficient rear center of the planform, so that the lift curve slope inceases sharply with this parameter.

For the triangular wing ($\xi = 0$), the formulas for the lift curve slope reduce to

$$C_{L_\alpha} = \frac{2\pi m}{\beta E'(m)}, \qquad m \leqq 1 \qquad (14\text{-}9)$$

[19] $E'(m)$ is the complete elliptic integral of the second kind with modulus $\sqrt{1-m^2}$.

and

$$C_{L_\alpha} = \frac{4}{\beta}, \qquad m \geq 1 \qquad (14\text{-}10)$$

Figure A,14l shows this variation, together with experimental data taken from various sources. We have already seen that the local pressures are accurately predicted by the theory when the leading edge is well within

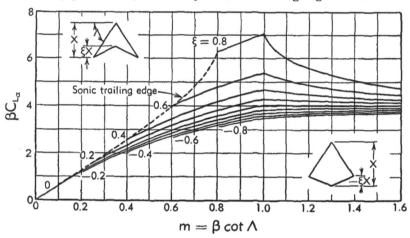

Fig. A,14j.   Variation of lift curve slope with Mach number, sweep angle, and trailing edge cutout for wings tapered to a point (supersonic trailing edges).

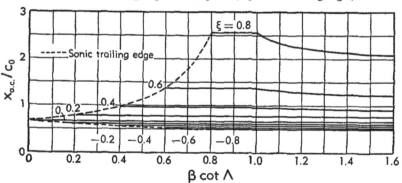

Fig. A,14k.   Variation of aerodynamic center location for the wings of Fig. A,14j.

the Mach cone from its apex. Accordingly, the total lift also follows the theoretical curve closely when the Mach number and/or aspect ratio is low. As the Mach line approaches the leading edge ($m \to 1$; $\beta R \to 4$), the experimental values appear to fall below the theoretical ones. The reason for this failure to realize the theoretical lift in this case is not understood at present.

In computing the drag due to lift, it is necessary, when m is less than 1, to take into account the suction force at the leading edge. From Eq. 13-34 we obtain as the strength of the leading edge singularity

$$C_\Delta = \lim_{x \to \frac{\beta y}{m}} u_\Delta \sqrt{x - \frac{\beta y}{m}} = u_0 \sqrt{\frac{x}{2}} \qquad (14\text{-}11)$$

with

$$u_0 = \frac{mU\alpha}{\beta E'(m)} \qquad (14\text{-}12)$$

Then substituting in Eq. 12-1 and integrating gives the thrust

$$T = \frac{\pi}{2} \rho_\infty U^2 \alpha^2 \frac{s^2}{[E'(m)]^2} \sqrt{1 - m^2} \qquad (14\text{-}13)$$

Subtracting this quantity, which depends only on the Mach number, sweep, and spanwise extent of the leading edge, from the product of the

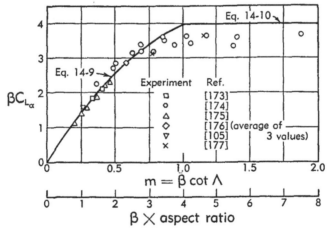

Fig. A,14l. Lift curve slope of triangular wings at supersonic speeds; theory and experiment.

lift and the angle of attack gives the drag due to angle of attack for any wing with straight subsonic leading edges, provided that there is no disturbance of the oncoming stream. The ratio C_D/C_L^2 for pointed-tip wings is plotted against m for various values of ξ in Fig. A,14m.

In Fig. A,14n, comparison is made of the experimental drag of two wings differing only in the rounding of the leading edge. The wings were of triangular planform, with the sweep and Mach number such that m had the value of 0.58. It is seen that the sharp-nosed airfoil showed a drag rise almost equal to the theoretical value for the wing in the absence of leading edge suction. A slight rounding of the nose reduced the drag to values approaching those predicted by the more complete theory. (The effect is of course not expected to be maintained beyond a small range of angle of attack.)

Wings of hexagonal planform with supersonic trailing edges. We will consider next wings having sweptback leading edges, blunt tips, and

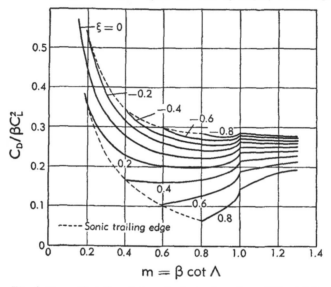

Fig. A,14m. Variation of drag rise factor for wings of Fig. A,14j.

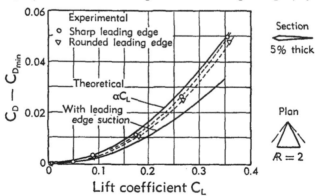

Fig. A,14n. Effect on drag rise of rounding the leading edge on a sweptback wing (Double wedge section rounded to radius of 0.25% chord.) $M_\infty = 1.53$. From [176].

trailing edges swept either forward or back, but not so much as to be subsonic. Except in the neighborhood of the tips, the lift distributions on such wings are the same as for the pointed wings (of the preceding discussion) having the same sweep of the leading and trailing edges. If the leading edge is supersonic, the Evvard-Krasilshchikova method gives the lift distribution within the tip Mach cones immediately (see Table A,13b). Formulas and numerical results obtained by this method for the loading and lift curve slopes of a large family of wing planforms are available in [*178*]. The pressure distributions are typified by that shown in Fig. A,14o. In the region between the leading edge and the Mach cones from the apex and tips, the loading is constant. The discontinuities in the planform outline appear in the loading as discontinuities in the pressure gradient at the Mach lines from the corners of the wing.

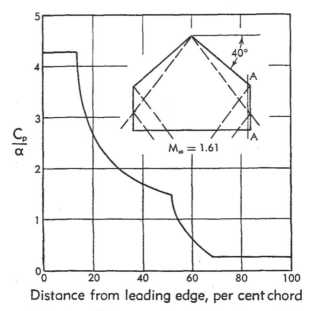

Fig. A,14o. Distribution of lifting pressure near the tip of a sweptback wing with supersonic leading and trailing edges and streamwise tips (section AA in sketch).

The lift distribution on wings with subsonic sweptback leading edges has been treated in considerable detail in connection with the lift-cancellation methods. Fig. A,14p shows the same wing as that of Fig. A,14o at a somewhat lower Mach number. The section lift distribution was calculated by means of Eq. 13-34 and 13-38a. Two basic differences are immediately noted: in addition to the appearance of the familiar leading edge suction peak, because of the now subsonic normal component of the oncoming stream, there also occurs a discontinuity in the *magnitude* of the pressure, rather than in the gradient, at the edge of the Mach cone from the tip. The latter effect is of course the consequence of the former and occurs whenever an abrupt change in conditions is introduced where the velocities are (theoretically) infinite.

The magnitude of the jump in the local lift at a point on the tip Mach line is readily calculated, being equal to

$$\Delta p = -\frac{\rho_\infty U u_0 \sqrt{b}}{\sqrt{(1+m)d}} \tag{14-14}$$

in which $2\rho_\infty u_0 U$ (Eq. 14-12) is the magnitude of the lifting pressure along the center line, m is β times the tangent of the semiapex angle, b is the

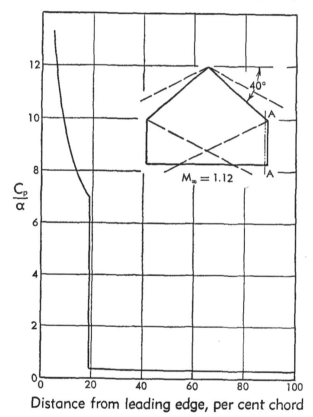

Fig. A,14p.   Lift distribution at same spanwise station of the wing shown in Fig. A,14o, at a Mach number such that the leading edge is subsonic.

wing span, and d the spanwise distance of the point from the tip. The jump, which always results in the loss of more than half the local lift on crossing the tip Mach line, is nicely illustrated in a set of isobar diagrams calculated by Gilles [179], two of which are reproduced in Fig. A,14q. It is interesting to note that the isobars in the tip regions are, for practical purposes, straight lines. Since these lines are also almost parallel to the stream, it will usually be sufficient merely to calculate the pressure at points along the tip Mach lines with and without the tip effect, and to assume the lift to be constant at the former values rearward from these points (see also Fig. A,14p).

At the lower Mach number of the two for which the isobars are shown, the tip Mach cones overlap. Superposition of the tip losses in the region of

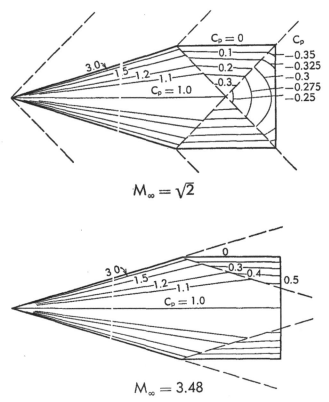

Fig. A,14q. Isobars calculated for a sweptback wing with subsonic leading edge and streamwise tips (from [179]).

overlapping results in negative lift. At still lower Mach numbers, reflection of the Mach waves in the side edges would restore some of this lift. However, it is clear that, particularly for low supersonic speeds, the rearward portion of such low aspect ratio wings is aerodynamically inefficient.[20]

The integrated lift of hexagonal wings with supersonic leading and trailing edges and streamwise tips is, in coefficient form,

[20] Note that, for $M_\infty = 1$, linear theory indicates *no* lift in this region.

$$C_L = \frac{2mR\alpha}{\pi a_t^2 \sqrt{m^2-1}} \left\{ m\, \frac{(m_t - a_t)^2}{m_t^2 - m^2} \left(\frac{m}{m_t} \cos^{-1}\frac{1}{m} - \frac{m_t}{m} \sqrt{\frac{m^2-1}{m_t^2-1}} \cos^{-1}\frac{1}{m_t} \right) \right.$$

$$+ \left(\frac{m_t}{m}\right)^{\frac{3}{2}} \frac{1}{\sqrt{m_t+1}} \left[\frac{(m+a_t)^2 \sqrt{m-1}}{m_t+m} \cos^{-1}\sqrt{\frac{(1+m)a_t}{m+a_t}} \right.$$

$$\left. - \frac{(m-a_t)^2 \sqrt{m+1}}{m_t-m} \cos^{-1}\sqrt{\frac{m(1-a_t)}{m+a_t}} \right]$$

$$\left. + 2m_t \frac{(m_t-a_t)^2}{m_t^2-m^2} \sqrt{\frac{m^2-1}{m_t^2-1}} \cos^{-1}\sqrt{\frac{m_t(1-a_t)}{m_t-a_t}} \right\} \qquad (14\text{-}15)$$

The symbols used in Eq. 14-15 are defined in Fig. A,14r. The formula presented covers only those cases in which the trailing edge does not

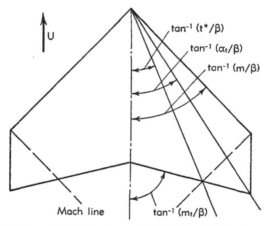

Fig. A,14r. Sketch showing symbols for Eq. 14-15 and 14-16.

change direction inside the tip Mach cones and $a_t \lessgtr 1$. Special forms for cases in which Eq. 14-15 becomes indeterminate (untapered wings, sonic edges, and straight trailing edges), as well as charts of computed lift curve slopes for a large range of parameters, are given in [178] and [180]. Span load distributions for hexagonal wings with supersonic leading edges are available in [181].

If the leading edge of the wing is subsonic, the tip correction to the lift cannot be integrated in terms of tabulated functions (see [155]). Calculated values are included in the charts of [180] and, when available, provide the most practical way of determining the lift on wings with supersonic trailing edges. The following approximate formula is based on the previously suggested assumption of uniform loading along each section downstream from the tip Mach line:

$$C_L = \frac{\alpha R}{E'(m)} \left\{ \frac{(1+m)^2(m_t-t^*)^2}{(m_t^2-m^2)(1+t^*)^2} \left[\frac{m_t}{\sqrt{m_t^2-m^2}} \left(\cos^{-1}\frac{m^2-m_t t^*}{m(m_t-t^*)} \right. \right. \right.$$

$$- \cos^{-1}\frac{m}{m_t}\Bigg) - \frac{\sqrt{m^2 - t^{*2}}}{m_t - t^*} + \frac{m}{m_t}\Bigg]$$

$$+ \frac{1 + m}{1 - m}\left[\frac{1}{\sqrt{1 - m^2}}\cos^{-1}\frac{t^* + m^2}{m(1 + t^*)} - \frac{\sqrt{m^2 - t^{*2}}}{1 + t^*}\right]$$

$$+ \frac{1 + m_t}{m_t(1 - m)^2}\left[\frac{\sqrt{m^2 - t^{*2}}}{1 + t^*}\left(2 + \frac{t^* + m^2}{1 + t^*}\right)\right.$$

$$\left. - \frac{1}{\sqrt{1 - m^2}}\left(m^2 + 2\frac{t^* + m^2}{1 + t^*}\right)\cos^{-1}\frac{t^* + m^2}{m(1 + t^*)}\right]$$

$$\left. - \frac{4(1 + m_t)\sqrt{2}}{3m_t\sqrt{m(1 + m)}}\left(\frac{m - t^*}{1 + t^*}\right)^{\frac{3}{2}}\right\} \quad (14\text{-}16)$$

The formula is derived for wings with supersonic trailing edges that do not change direction within the tip Mach cones. The symbols are defined in Fig. A,14r.

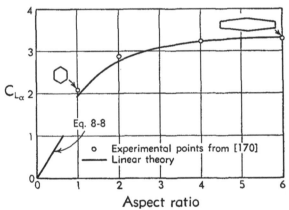

Fig. A,14s. Variation of lift curve slope with aspect ratio for tapered wings with unswept 50 per cent chord line; $\lambda = \frac{1}{2}$, $M_\infty = 1.53$.

A typical variation of lift curve slope with aspect ratio is shown in Fig. A,14s, taken from [170]. The curve is for wings with fore-and-aft symmetry (unswept midchord line) and a fixed taper ratio of $\frac{1}{2}$, so that there is considerable variation of the leading edge sweep. Fixing the leading edge sweep and reducing the aspect ratio by cutting off area at the tips give a slightly sharper variation of lift curve slope. When the fifty per cent chord line is unswept there is very little variation of lift curve slope with taper ratio (Fig. A,14t), but it will be seen later (Fig. A,14bb) that this is not generally the case.

Since, on the wings being considered in this section, the leading edge is always outside the zone of influence of the tips, the drag due to angle of attack is calculated exactly as for the pointed wings. If the leading edge is subsonic, the reduction in drag is given by Eq. 14-13; otherwise the drag is merely the product of the lift and the angle of attack. Span load distributions for these wings can be found in [182].

Sweptback wings. The preceding paragraphs dealt with hexagonal wings with supersonic trailing edges. At lower speeds the trailing edges of these wings will be subsonic and, as previously pointed out, the mathematical treatment of the problem remains incomplete. Because of the special interest attached to highly swept wings, the inquiry into their characteristics has nevertheless been pushed as far as possible. As a result, the theoretical distribution of lift can be ascertained with sufficient accuracy and completeness for any practical purpose when the planform and Mach number are such that the Mach lines from the trailing edge apex intersect the tips, but not the leading edge (Fig. A,13j). Although the

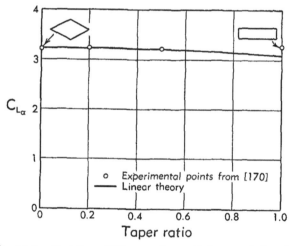

Fig. A,14t. Variation of lift curve slope with taper ratio for wings with unswept 50 per cent chord line; $\mathcal{R} = 4$, $M_\infty = 1.53$.

multiple reflections of tip and trailing edge Mach lines make it impossible to complete the solution, the loading ahead of the first reflections is closely approximated by combining Eq. 13-34 and 13-38a with Eq. 13-40. In [155] can be found simple approximations to the effects of the first reflections of the Mach lines. Farther to the rear the exact solution would appear to be only of academic interest, in view of the limitations of the theory being employed.

A typical load distribution in the cases described is shown in Fig. A,14u. The distribution is given in more detail for the section at 75 per cent semispan (Fig. A,14v) in order to show the magnitude of the various edge effects. Component 1 corresponds to the first term in Eq. 13-40, the effect of canceling a uniform lift equal to the lift at the root section, all along the trailing edge. Component 2 is the approximate effect of

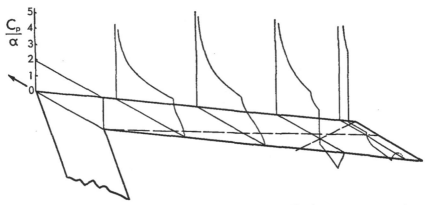

Fig. A,14u. Isometric sketch of the theoretical load distribution on an untapered, sweptback wing with subsonic leading and trailing edges. $\beta \cot \Lambda = 0.6$, $\beta Æ = 1.92$.

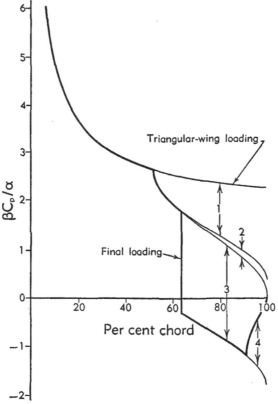

Fig. A,14v. Details of load distribution calculated by linearized theory for section at 75 per cent semispan of wing of the preceding figure.

canceling the remaining lift behind the wing. Component 3 is the primary tip effect (Eq. 13-38a), and component 4 the "reflection" of this effect in the trailing edge. It may be seen that the major portion of the trailing edge effect is given by a single conical field (solution 18 of Table A,13a). The corresponding decrement in total lift is readily found:

$$\frac{\Delta L}{\frac{1}{2}\rho_\infty U^2} = -\frac{4\alpha s^2 m}{m_t E'(m)}\left[1 - \frac{\pi/2}{K'(m_t)}\right] \tag{14-17}$$

where E' and K' are the complete elliptic integrals with complementary moduli. Combined with Eq. 14-16, Eq. 14-17 provides a fairly good estimate of the lift coefficient for a tapered sweptback wing with trailing edge Mach cone intersecting the wing tips. For more complete formulas, see [155].

If the trailing edge Mach lines cross the leading edge as in Fig. A,13k, one must begin to look toward the asymptotic cases. Two limiting solutions are available, one for Mach numbers approaching unity, and the other for aspect ratios approaching infinity. Calculated lift distributions show (Fig. A,14w) that, behind the reflection of the trailing edge Mach wave at the leading edge, the loading already resembles very closely that on the two-dimensional flat plate in incompressible flow. If the section is taken spanwise (the chordwise distributions shown in Fig. A,14w may then be considered to be crossplots from spanwise section curves), and a correction factor is introduced to make the two loadings agree in the strength of the singularity at the leading edge, it is found that the agreement is good everywhere; that is, on untapered wings extending infinitely far downstream, the loading approaches the two-dimensional essentially as the strength of the leading edge singularity. Even the tapered wing (Fig. A,14x) may be treated in this way, if the taper is not extreme.

Also shown in Fig. A,14w are lift distributions computed by slender wing theory (Art. 8). In the latter theory, the flow is assumed from the start to be two-dimensional in spanwise sections. Since both branches of the wing are taken into account, the expression for the lift distribution has a more complicated form than for the single plate. However, the calculations made indicate very little effect of the presence of the far wing on the shape of the lift distribution, so that the simpler expression is probably adequate. There remains only the problem of determining the correction factor to be applied.

Ahead of the trailing edge Mach cone, the strength of the leading edge singularity is that on the triangular wing having the same sweep of the leading edge (Eq. 14-11). Behind the shock wave from the trailing edge, the flow is of course modified. The effect on the leading edge singularity may be calculated for some distance downstream of the shock by formulas given in [155]. A good approximation to the modified coefficient is

$$C_\Delta(x) + \Delta C(x) = u_0\left[\frac{1}{\sqrt{2}} - \frac{4mK(k)Z(\psi, k)}{\pi m_t K'(m_t)\sqrt{1 + m}}\right]\sqrt{x} \tag{14-18}$$

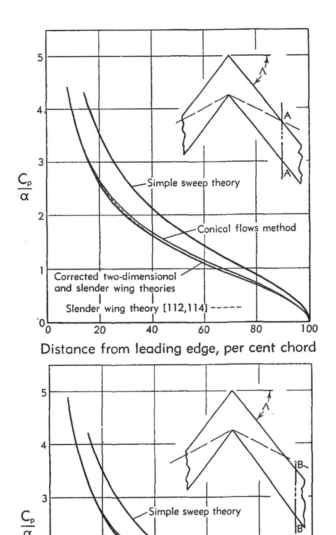

Fig. A,14w. Theoretical lift distributions on an untapered, sweptback wing,
$\beta \cot \Lambda = 0.4$, on sections entirely within the trailing edge Mach cone.

in which m_t is β times the cotangent of the angle of sweep of the trailing edge, and, in the elliptic functions K and Z [151],

$$k = \sqrt{\frac{2(y_2 - m_t y_1)}{(1 - m_t)(y_1 + y_2)}}, \qquad \psi = \sin^{-1}\sqrt{\frac{y_1 + y_2}{2y_1}}$$

(see Fig. A,14y for y_1 and y_2). Expressing the two-dimensional subsonic

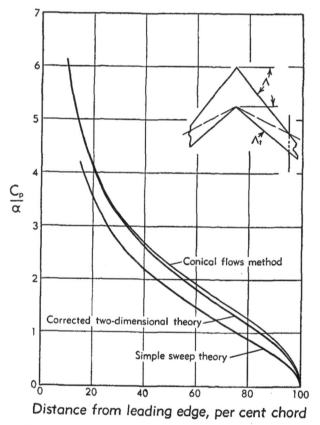

Fig. A,14x. Theoretical lift distribution on a tapered, sweptback wing; $\beta \cot \Lambda = 0.4$, $\beta \cot \Lambda_t = 0.6$; section entirely within the trailing edge Mach cone.

loading (formula 3, Table A,2) in the same form and adjusting the coefficient to give the above leading edge value, we obtain

$$u(x,y) = [C_\Lambda(x) + \Delta C(x)] \sqrt{\frac{m(y - y_2)}{\beta(y_1 - y_2)(y_1 - y)}} \qquad (14\text{-}19)$$

With this information, the investigation of sweptback wings can be extended a little further in the direction of increasing aspect ratio. Approximate formulas [*155*] for the loss of lift in the tip region and for the total lift may be derived. Since the strength of the leading edge singularity is known, the leading edge thrust and hence the drag due to lift may be calculated. However, if the wing shown in Fig. A,13k were of somewhat greater aspect ratio, or if the Mach number were slightly lower so that the relative sweep of the wing were greater, the second reflection of the trailing edge Mach wave, now shown intersecting the tip, would again cross the leading edge, and further modification of the leading edge

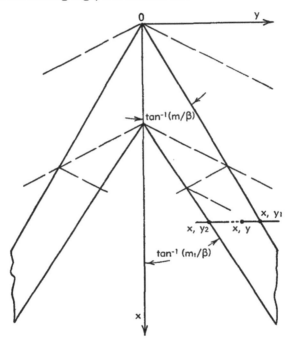

Fig. A,14y. Sketch showing symbols for Eq. 14-18 and 14-19.

singularity would take place. The exact manner in which the successive modifications cause the flow to approach the two-dimensional is not known at present. The slender wing (or $M_\infty \rightarrow 1$) theory, which appears to predict the pressure quite well in the range for which we have the exact solution, unfortunately gives values of the leading edge coefficient too low by the factor $\sqrt{1 - m^2}$ as $x \rightarrow \infty$. Thus there remains, at the high aspect ratio end of the range, a considerable gap in the theory, as shown by Fig. A,14z and A,14aa, where the lift curve slope and the drag due to lift are plotted against aspect ratio for untapered wings. In these figures, the calculations for $m = 0.2$ and 0.4, from [*155*], have been carried out as far as the current state of the theory permits. The asymptotic values for infinite aspect ratio—that is, $C_D = 0$ and, from simple sweep theory,

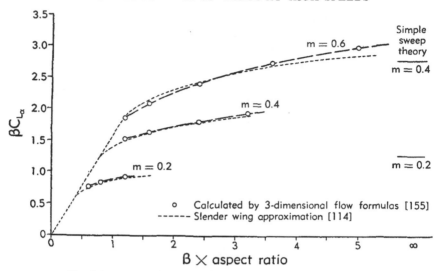

Fig. A,14z.   Variation of lift curve slope of sweptback, untapered wings with aspect ratio and Mach number.

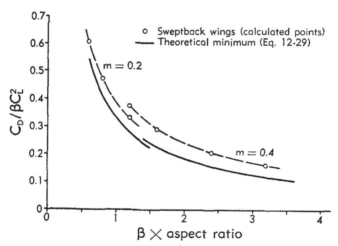

Fig. A,14aa.   Drag rise factor for sweptback, untapered wings, as a function of Mach number and aspect ratio.

$\beta C_{L_\alpha} = 2\pi m/\sqrt{1-m^2}$—are also shown, as well as the slender wing theory values for C_{L_α}. The latter appear to be in remarkable agreement with the results of the three-dimensional theory over a considerable range of aspect ratio, Mach number, and sweep angle. Results for two tapered wings, presented in [155], were likewise duplicated within the accuracy of the computations by slender wing theory. However, it should be recognized that these coincidences are not to be generally expected, but are the result of compensating errors, as indicated by the deviation of the curve of C_{L_α} for triangular wings (Fig. A,14l) from the straight line $C_L = \frac{1}{2}\pi R$.

In the low aspect ratio range, interference between the disturbance fields of the two tips further complicates the calculations and puts a practical limit on the use of the three-dimensional theories. If the leading edge of the wing is highly swept (relative to the Mach lines), the slender wing theories of [110–114] will provide solutions of reasonable accuracy. The chief discrepancies will occur in the tip regions, where the slender wing theory result of zero lift is approached by the three-dimensional solution only as M_∞ approaches 1. Thus, if such regions form a large part of the wing planform, or if the trailing edge is swept forward, as in a diamond, a more accurate procedure such as that of [115] will be required.

The dotted curves in Fig. A,14z, being for untapered wings, were computed using Mirels' numerical results [114] (as were the pressure distributions of Fig. A,14x). Results for other taper ratios have been given by Mangler [112] and by Legendre and Eichelbrenner [113]. Legendre has fitted the numerical results with a simple formula which, if λ is the taper ratio and μ is the sweep ratio m/m_t, may be written

$$C_{L_\alpha} = \frac{\pi}{2} R \frac{(1 + \lambda)(1 - \mu)}{1 - \lambda\mu} \left[1 - 0.004(1 + 6\lambda^2)\frac{\mu - \lambda}{1 - \mu} \right], \qquad \lambda < \mu \leqq 1$$

$$(14\text{-}20a)$$

For untapered wings ($\lambda = \mu = 1$) the formula becomes[21]

$$C_{L_\alpha} = \frac{2\pi R}{R + 2\cot\Lambda}(1.028\cot\Lambda - 0.014R) \qquad (14\text{-}20b)$$

$$R \geqq 2\cot\Lambda$$

Comparison with the results of [114], shows Eq. 14-19b to hold through $R = 6\cot\Lambda$, with a deviation of about 3 per cent at $R = 8\cot\Lambda$. The cases excluded from Eq. 14-20a and 14-20b are those in which the trailing edge lies entirely in the nonlifting region of the wing, when the formula is simply (Art. 8)

$$C_{L_\alpha} = \frac{\pi}{2} R$$

Making use of slender wing values (for $M_\infty = 1$) and conical flow formulas, Eichelbrenner [133] has computed charts from which the lift

[21] The slender wing formulas, which are in a sense derived for sonic speed and do not show the effect of Mach number, are generalized to Mach numbers slightly different from 1 by writing βC_{L_α} for the left-hand side and substituting for $\cot\Lambda$ and R, $m = \beta\cot\Lambda$ and βR, the similarity parameters suggested by the Prandtl-Glauert transformation.

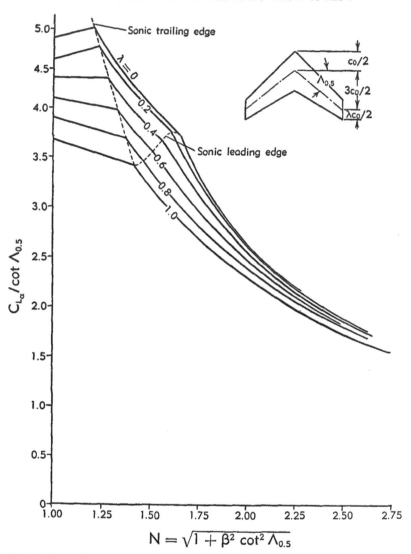

$$N = \sqrt{1 + \beta^2 \cot^2 \Lambda_{0.5}}$$

Fig. A,14bb. Chart showing effect of taper on lift curve slope of a family of swept-back wings with fixed depth-to-chord ratio as defined in sketch. One of a set of design charts given in [183].

curve slope for a given wing may be determined over the whole Mach number range. In the form shown in Fig. A,14bb, the charts show the effect of taper ratio and Mach number for selected values of a planform parameter involving the aspect ratio and sweep. The departure from the horizontal near $M_\infty = 1$ indicates the error in using slender wing theory for any particular case. However, the compressed scales in M_∞ should be noted.

The drag, according to slender wing theory, is computed from the span loading, plus a term of higher order in β, representing the wave drag, (see Art. 12). From [*113*], the span loading (Fig. A,14cc) is seen to have the form associated with the sweptback wing in subsonic flight, as contrasted with the elliptical span loading predicted by the same theory for wings of convex planform. The curve plotted in Fig. A,14aa as the theo-

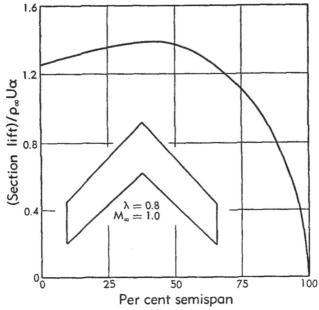

Fig. A,14cc.   Theoretical span loading on tapered sweptback wing at a Mach number of 1 [*113*].

retical minimum is based on an elliptical span loading, but includes the wave drag.

A,15. Wave Drag and Pressure Distribution Due to Thickness.

General formulas for wave drag due to thickness. To obtain general expressions for the wave drag due to thickness we make use of the fact that the drag is unchanged by a reversal of the direction of motion (see VI,D,31). After superimposing the perturbation velocities in forward and reversed motion, the drag is obtained from the distribution of pressure in the combined field.

The effect of a symmetrical (nonlifting) distribution of thickness may be obtained by superimposing fundamental solutions of the form which corresponds to the combined disturbance field of a source. The local source strength is proportional to the vertical perturbation velocity w, which is related to the prescribed thickness distribution $t(x_1, y_1)$ by the equation

$$w = \frac{1}{2}\frac{\partial t}{\partial x_1} U \tag{15-1}$$

By following steps similar to those outlined in the lifting case we obtain for the longitudinal and vertical perturbation velocities in the combined flow field:

$$\bar{u} = -\frac{1}{4\pi^2}\int_C \iint_S \frac{w(x_1, y_1)dx_1dy_1}{[\alpha(x - x_1) - \beta(y - y_1) - \gamma z]^2}\frac{\alpha d\lambda}{i\lambda} \tag{15-2}$$

$$\bar{w} = \frac{1}{4\pi^2}\int_C \iint_S \frac{w(x_1, y_1)dx_1dy_1}{[\alpha(x - x_1) - \beta(y - y_1) - \gamma z]^2}\frac{\gamma d\lambda}{i\lambda} \tag{15-3}$$

The value of $\bar{w}$ should of course vanish at each point of the given planform S. The drag will be given by

$$D = \rho_\infty \iint_S \bar{u}w dx dy \tag{15-4}$$

Eq. 15-3 leads directly to the concept of equivalent linear source distributions originally introduced by Hayes [44].

Following the steps outlined in Art. 12, we introduce the variables X, X_1, Y_1, and obtain for Eq. 15-2

$$\bar{u} = -\frac{1}{4\pi^2}\int_C \iint_S \frac{w(X_1, Y_1)dX_1dY_1}{(X - X_1)^2}\frac{\alpha d\lambda}{i\lambda} \tag{15-5}$$

For $\lambda = e^{i\theta}$, the lines $X = $ const correspond to the intercepts of a system of plane waves at 45° to the x axis ($M_\infty = \sqrt{2}$) and at an angle θ in the y, z plane. Eq. 15-5 then yields the three-dimensional disturbance field by the superposition of elementary plane waves at varying angles θ around the x axis. Inspection of Eq. 15-5 shows that at a particular value of θ the influence of an element of thickness w depends only on the relative position of its intersecting plane wave front, and is independent of its position Y_1 in the plane of the wave. An integration over Y_1 thus corresponds to the collection of sources at a single point in each plane wave, and, for the whole system, results in an equivalent linear source distribution of strength

$$\sigma'(X_1, \theta) = \frac{2}{U}\int w(X_1, Y_1)dY_1 \tag{15-6}$$

for each angle θ (see Fig. A,15a). Then the double integral over S in Eq.

15-5 may be written

$$\frac{U}{2} \int_{-X_s}^{+X_s} \frac{\sigma'(X_1, \theta) dX_1}{(X - X_1)^2} \tag{15-7}$$

which is the form of the integral for the induced downwash in the Prandtl wing theory. Following a procedure employed in that connection [*140*, pp. 135–140], we may expand the function $\sigma'(X, \theta)$ in the Fourier series

$$\sigma'(X, \theta) = \sum_{1}^{\infty} A_n \sin n\varphi \tag{15-8}$$

where $\varphi = \cos^{-1}(X/X_s)$ and the coefficients A_n are functions of θ. Then, as in lifting line theory, we may introduce Eq. 15-8 into 15-7 and obtain

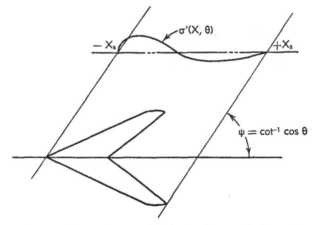

Fig. A,15a. Projected source distribution for a particular angle θ.

for the drag of each elementary linear source distribution

$$\frac{dD}{d\theta} = \int_{-X_s}^{X_s} \int_{-X_s}^{X_s} \frac{\sigma'(X, \theta)\sigma'(X_1, \theta) dX dX_1}{(X - X_1)^2} = \frac{\pi^2}{2} \sum_{1}^{\infty} n[A_n(\theta)]^2 \tag{15-9}$$

Finally after replacing $\alpha d\lambda / i\lambda$ by $d\theta$ in Eq. 15-5 we obtain

$$D = \frac{\rho_\infty U^2}{8} \int_0^\pi \sum_{1}^{\infty} n[A_n(\theta)]^2 d\theta \tag{15-10}$$

The coefficients A_n may be found from Eq. 15-8. Integration shows that the coefficient A_1 is given by

$$A_1 = \frac{2}{\pi} \frac{\sigma(X_s)}{X_s} \tag{15-11}$$

where X_s is of course a function of θ and the quantity

$$\sigma(X_s) = \int_{-X_s}^{X_s} \sigma' dX$$

is the "base area" of the wing. Hence A_1 will be zero if the wing is a finite closed body. Assuming $A_1 = 0$, we obtain for A_2

$$A_2 = -\frac{4}{\pi} \frac{\text{volume}}{X_s^2} \tag{15-12}$$

Higher terms of the series contribute nothing to the volume, but do contribute to the drag. We may thus obtain a lower bound for the drag with a given volume by suppressing higher terms of the series.

Relation of wave drag to area distribution. The foregoing equations represent the wave drag of a planar source distribution in terms of a series of equivalent linear source distributions. Thus, by making use of the well-known relation between source strength and cross-sectional area, the drag associated with the thickness of a wing may be obtained as the average of the values for a series of equivalent bodies of revolution.

As the Mach number approaches 1 the planes $X = $ const approach a single set of planes perpendicular to the flight direction. As remarked by Hayes [44], the drag of a wing, or of a wing-body system, then approaches the drag of a single equivalent body of revolution having the same longitudinal distribution of area.

It is interesting to note that this formal result of linear theory has been verified in a striking way by recent developments in transonic theory and experiment. On the theoretical side, Oswatitsch [184] has shown that the nonlinear transonic potential field of a slender wing-body combination approaches, with increasing radial distance, that of an equivalent body of revolution. In the vicinity of the wing or body the potential fields differ by a two-dimensional, laterally incompressible flow potential, as in the theory of Jones [105] or Ward [185].

Experiments made by Whitcomb [186] at transonic speeds show that the incremental drag of a wing-body combination follows rather closely the drag of the equivalent body of revolution, and furthermore that the drag of the complete airplane can be reduced considerably by re-shaping the fuselage so that the whole system presents a smooth area distribution along the longitudinal axis.

Wing shapes of minimum drag. Examples of wing shapes for which the wave drag due to thickness is a minimum are naturally of considerable interest to the wing designer. Such examples provide an indication of efficient wing shapes, and at the same time yield approximate engineering formulas for the drag of a variety of shapes which do not differ too greatly from the optimum.

The term minimum drag as used herein refers of course to the drag arising from pressure forces, exclusive of the friction. In practice, a balance must be achieved between the pressure drag and the friction drag. For present purposes, the friction drag is taken to be proportional to the wetted area of the wing surface. With a wing of given depth or thickness, increasing the chord diminishes the slope of the surface, and hence, the pressure drag; but the increased area results in a greater friction drag. Therefore the problem of minimizing the pressure drag for a wing having the smallest possible area is of interest.

Conditions for the drag to have the minimum value under certain auxiliary assumptions may be stated in terms of the pressure distribution in the combined flow field, given by

$$\overline{\Delta p} = \rho_\infty \bar{u} U \qquad (15\text{-}13)$$

The demonstration of these conditions involves a repetition of the reasoning employed in connection with the minimum drag of lifting surfaces (Art. 6 and [187]).

If the volume of the wing is specified, then the thickness distribution having the smallest wave drag must yield

$$\frac{\partial \overline{\Delta p}}{\partial x} \quad \text{or} \quad \frac{\partial \bar{u}}{\partial x} = \text{const} \qquad (15\text{-}14)$$

over the entire planform S. Since the pressure gradient $\partial \overline{\Delta p}/\partial x$ is the drag per unit volume, minimum drag requires that the drag per unit volume be constant.

Similarly, if a line is drawn on the wing planform and if the thickness distribution along this line is required to have a specified projected or frontal area, then it is found that $\overline{\Delta p}$ must have the same value at all points of the planform ahead of this line and again a constant value (perhaps different) at all points behind this line. In this case it will be observed that the drag per unit frontal area is constant in the superimposed disturbance field.

If the auxiliary assumptions are made more restrictive, then the condition for the drag to be a minimum becomes, of necessity, less restrictive. Thus, if the volume of the wing is specified but all sections along the span are required to have the same shape, then the streamwise gradient of the pressure must be permitted to vary along the span. The condition for minimum drag is then

$$\frac{\partial}{\partial x} \int_S \bar{u}\, dy = \text{const} \qquad (15\text{-}15)$$

Minimum wave drag for elliptic planform. If the planform of the wing is elliptical, the formulas (15-8) and (15-10) for the drag are considerably simplified. In particular, if the sections are assumed to be parabolic arcs such that

$$\frac{\partial t}{\partial x} = \text{const} \cdot x$$

over the whole planform, then the equivalent source distribution will be given by Eq. 15-18, with $n = 2$. This particular thickness distribution— which gives what might be termed an "elliptic lens"—thus yields the minimum drag consistent with a given volume. A similar parabolic biconvex section shape also yields the minimum drag in two-dimensional flow. With the elliptical planform, however, the thickness-chord ratio becomes smaller near the tips. The drag coefficient of the elliptic lens at $M_\infty = \sqrt{2}$ is given by

$$C_{D_{\min}} = \frac{t_{\max}^2}{a^2} \frac{1}{\sqrt{1 + \frac{a^2}{b^2}}} \left(1 + \frac{\frac{a^2}{b^2}}{1 + \frac{a^2}{b^2}} \right) \tag{15-16}$$

where a and b are the semiaxes of the ellipse.

Fig. A,15b shows the variation of drag coefficient with Mach number for elliptic wings of three different axis ratios. The values were obtained from Eq. 15-16 after applying the Prandtl-Glauert transformation. With the lowest aspect ratio the variation of drag with Mach number is almost completely suppressed.

It is of interest to compare the foregoing results with the results for a body of revolution designed to contain the same volume within the same specified length. The shape of the body and the minimum value of the drag coefficient in this case have been worked out by Sears [188] and Haack [189]. It may be imagined that the elliptic figure is derived by flattening the body of revolution. At first the flattening produces only a second order effect on the drag, but eventually the wave drag diminishes appreciably. If the body is flattened into a circular planform the wave drag (at $M_\infty = \sqrt{2}$) for the given volume and length is diminished to one half. Of course the wetted area and the skin friction are greatly increased by this process.

The drag for a given volume could be further reduced by increasing the value of X_s. Since X_s is one half the projected maximum dimension of the wing planform, this process would involve spreading the volume out over a larger area. However, since the projections involve only those angles lying between the forward and rearward Mach characteristic angles, it is clear that no value of $X_s(\theta)$ can be increased by additions to the planform if such additions are restricted to the interior of the characteristic envelope of the wing (see Fig. A,15c). Thus the drag of the elliptic lens, or of any system of sources for which Eq. 15-10 holds, cannot be diminished by any redistribution of volume elements within its characteristic envelope.

Calculated wave drags. It is of course possible to calculate the wave drag of a particular wing by the integration of the theoretical pressures on the surface. The determination of these pressures, in the case of a thin symmetrical airfoil at zero lift, is a problem that lends itself readily

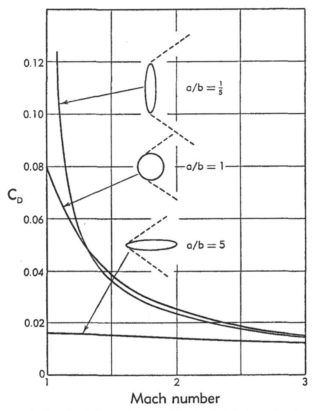

Fig. A,15b. Variation of drag coefficient with Mach number for elliptic wings with biconvex sections, $t/c = 0.1$, at zero lift.

to analysis by the linearized theory. For many simple wing shapes the entire process can be carried out analytically. For example, complete results for wings of triangular planform with double-wedge sections were early given by Puckett and Stewart [172]. Swanson and Harmon [190,191] present formulas and charts for untapered wings of biconvex section. Bismut [192] treats the "swallowtail" wing, including results for wings with parabolic arc section, Chang [193] and Nielsen [194] consider tapered wings with diamond profile (that is, maximum thickness at 50 per

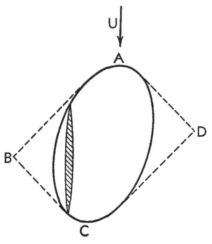

Fig. A,15c. Ideal distribution of thickness for area *ABCD*, given volume.

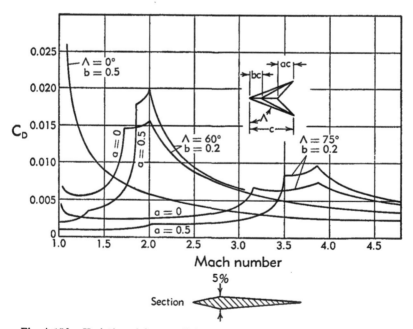

Fig. A,15d. Variation of drag coefficient with Mach number for "swallowtail" wings with double-wedge sections, at zero lift [*172*].

cent chord), and Margolis [*195,196,197*] covers the double-wedge profile with varying location of the maximum thickness line. Most of these results have been summarized by Lawrence [*198*] in the form of design charts.

Some typical results of these calculations are reproduced in Fig. A,15d, A,15e, and A,15f. A characteristic feature of all such calculations for wings of rectilinear planform is the occurrence of sharp peaks in the variation of wave drag with Mach number. Such peaks arise when the Mach lines coincide with bends in the stream surface, which ordinarily

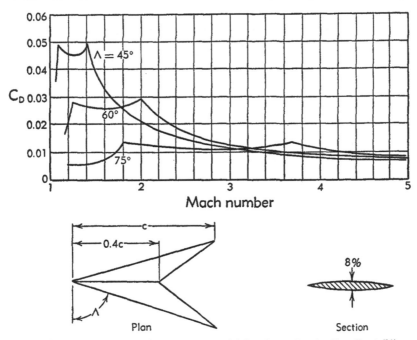

Fig. A,15e. Variation of drag coefficient with Mach number for "swallowtail" wings with biconvex sections, at zero lift [*192*].

occur at the leading and trailing edges of the wing. The linearized theory, of course, displays the Mach lines in a perfectly rectilinear pattern, undistorted by superimposed variations in the local relative velocity. In the vicinity of a leading edge or a trailing edge the disturbance velocity becomes very large—of the order of the stream velocity—so that this coincidence of the Mach lines can hardly be expected to occur in practice. Furthermore, the effect of friction will lead to a smoothing of the effective contour of the airfoil at the trailing edge, so that a sharp bend of the stream surface in this region will not occur.

A situation similar to the foregoing arises in the prediction of the wave resistance of ships [199]. Here, calculations based on a linearized theory, disregarding friction, show a succession of peaks in the variation of wave resistance with speed, the peaks occurring when multiples of the wave length coincide with the length of the ship. By supposing a cusp-shaped extension added to the ship, smoothing out the lines of the stern to represent the probable effect of the boundary layer, the peaks in the wave resistance can be eliminated from the calculations and the results can be brought into better agreement with experimental values [199].

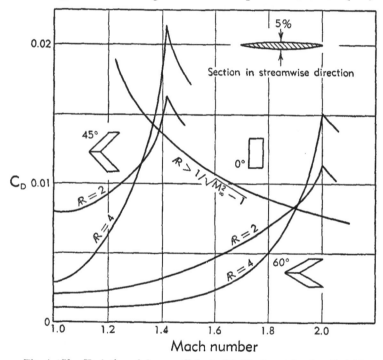

Fig. A,15f. Variation of drag coefficient with Mach number for straight and sweptback wings with biconvex sections, at zero lift [190].

Calculated pressure distributions for sweptback wings. Perhaps the most direct approach to the problem of determining the pressure distribution over symmetrical nonlifting airfoils is the method of source distributions (see Art. 13) suggested by von Kármán in his 1935 Volta Congress paper [200]. For airfoils composed of cylindrical or conical surfaces it is convenient to consider the sources to be distributed along the generating elements of the surface. If the source strength is held constant along each line, the distribution of source strength can then be represented in terms of a single variable. This simplification makes possible the treatment of fairly general section shapes.

A similar procedure was employed in Art. 9 in treating the corresponding problem in subsonic flow. The application to supersonic flow is found in [*127*]. As in Art. 9, the fundamental solution (10-10) may be used to represent the longitudinal velocity component u rather than the velocity potential φ. Then to get the effect of a line of sources of uniform density along the x axis we merely integrate the expression for the fundamental solution:

$$\frac{u_0}{U}(x, y, z) = -\int_0^\xi \frac{dx_1}{\sqrt{(x - x_1)^2 - y^2 - z^2}}$$

$$= -\cosh^{-1}\frac{x}{\sqrt{y^2 + z^2}} \tag{15-17}$$

letting ξ represent the location of the last source whose Mach cone includes the points x, y, z, i.e.

$$\xi = x - \sqrt{x^2 + z^2}$$

Such a distribution of sources along the x axis satisfies the boundary condition for a circular cone, and corresponds to the solution for that case given by von Kármán and Moore [*201*]. The solution for an oblique line source may be derived by integrating to obtain the effect of a row of sources along the oblique line $y = mx$. It is more convenient, however, to make use of the transformation to oblique coordinates discussed in Art. 13, that is,

$$\left.\begin{aligned} x' &= x - my \\ y' &= y - mx \\ z' &= \sqrt{1 - m^2}\, z \end{aligned}\right\} \tag{15-18}$$

The expression for the oblique line source then becomes, for $m < 1.0$,

$$\frac{u_0}{U} = -\cosh^{-1}\frac{x'}{\sqrt{y'^2 + z'^2}} \tag{15-19}$$

and for $m > 1.0$,

$$\frac{u_0}{U} = -\cos^{-1}\frac{x'}{\sqrt{y'^2 + z'^2}} \tag{15-20}$$

Since u is proportional to the pressure through the relation

$$\frac{\Delta p}{\frac{1}{2}\rho_\infty U^2} = -2\frac{u}{U}$$

these equations describe the pressure field associated with the source line. The vertical velocity w near $z = 0$ determines the shape of the boundary and may be obtained by integrating u with respect to x and then differentiating the resulting velocity potential with respect to z. Thus

$$w = \frac{\partial \varphi}{\partial z} = \frac{\partial}{\partial z} \int u\, dx \qquad (15\text{-}21)$$

When this operation is carried out for either Eq. 15-19 or 15-20 it is found that w near $z = 0$ is zero everywhere except over the radial sector of the plane in which $y' < 0 < y$ (see Fig. A,13c). Over this area w has a constant value, given by

$$\frac{w_0}{U} = \pm \pi \frac{\sqrt{1 - m^2}}{|m|} \qquad (15\text{-}22)$$

The constant value of w/U implies a constant value of $\partial z / \partial x$, which in this case takes on equal and opposite values on the upper and lower sides of the chord plane. The solutions (15-19) and (15-20) thus satisfy the boundary condition for a thin oblique wedgelike body having its leading edge along the line $y' = 0$. The relation between the thickness and the magnitude of the pressure disturbance may be obtained by introducing the relation

$$\frac{\partial z}{\partial x} = \frac{w}{U}$$

together with Eq. 15-22. The result is

$$\frac{\Delta p}{\frac{1}{2}\rho_\infty U^2} = \begin{cases} \dfrac{2}{\pi} \dfrac{\partial z}{\partial x} \dfrac{m}{\sqrt{1 - m^2}} \cosh^{-1} \dfrac{x'}{\sqrt{y'^2 + z'^2}}; & \text{for } |m| < 1 \\[3mm] \dfrac{2}{\pi} \dfrac{\partial z}{\partial x} \dfrac{m}{\sqrt{m^2 - 1}} \cos^{-1} \dfrac{x'}{\sqrt{y'^2 + z'^2}}; & \text{for } |m| > 1 \end{cases} \qquad (15\text{-}23)$$

For a bilaterally symmetrical wedge, we may superimpose the foregoing solution and its mirror image in the x axis, since in each solution w is zero in the portion of the x, y plane occupied by the other. The terms for the left-hand wedge are obtained by replacing m by $-m$ in Eq. 15-18 and 15-23.

Since the flows produced by the wedge-shaped figures are conical in form, they may also be derived by the conical flow method described earlier. To satisfy the boundary condition for a bilaterally symmetrical wedge we write

$$w_0 = \text{R.P.} \frac{i}{\pi} \ln \frac{\tau + m}{\tau - m} \qquad (15\text{-}24)$$

where τ is the conical variable defined by Eq. 13-1 and 13-2. The real part of this function (Fig. A,13b) shows the desired behavior, yielding the discontinuous value

$$w = \pm 1$$

between the points $\tau = -m$ and $+m$.

The corresponding values of u_0 and v_0 are found by integrating the relations (13-3) and (13-4). The integration yields, for $m < 1.0$,

$$u_0 = \text{R.P.} \frac{-m}{\pi \sqrt{1-m^2}} \left(\cosh^{-1} \frac{1-m\tau}{m-\tau} + \cosh^{-1} \frac{1+m\tau}{m+\tau} \right) \quad (15\text{-}25)$$

$$v_0 = \text{R.P.} \frac{1}{\pi \sqrt{1-m^2}} \left(\cosh^{-1} \frac{1-m\tau}{m-\tau} - \cosh^{-1} \frac{1+m\tau}{m+\tau} \right) \quad (15\text{-}26)$$

The expression for u_0 is seen to correspond to the superposition of two (right- and left-hand) disturbance fields of the form 5, Table A,13a; the

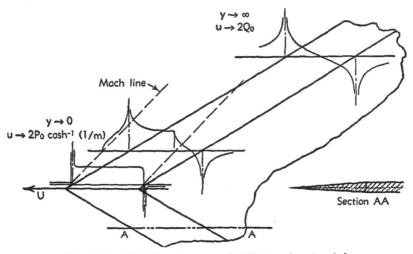

Fig. A,15g. Variation of pressure distribution along beveled edge. $M_\infty = \sqrt{2}$; $m = \tan 30°$ [*127*].

latter can, in turn, be shown to reduce to Eq. 15-19, except for the constant coefficient. The corresponding expression for $m > 1.0$ is solution 3 of Table A,13a.

Airfoils of rectilinear planform bounded by a finite number of sloping plane surfaces, can be obtained by superimposing a finite number of solutions of the form (15-19) or (15-20). This process is described in detail in [*127*] and follows essentially the procedure described in Art. 9.

Fig. A,15g illustrates the result of superimposing two solutions of the form (15-25) with the origins displaced along the x axis. The figure obtained is the beveled leading edge of a plate of uniform thickness. The sweep of the leading edge in this example is 60° while the Mach number is $\sqrt{2}$, so that the normal component of the Mach number is less than 1.0. As shown in the figure, the pressure distribution over the sections approaches the subsonic form (see formula 19 of Table A,2) with increasing distance from the root section. At the root section the pressure is constant along the chord, as given by the Ackeret theory for a straight wing with wedge sections. Inspection of Eq. 15-25 at the limit $y = 0$ shows that the effect of sweep is to reduce the pressures at the root section by the factor

$$\frac{2}{\pi} \frac{m}{\sqrt{1 - m^2}} \cosh^{-1} \frac{1}{m} \qquad (15\text{-}27)$$

relative to the value given by the Ackeret theory.

From the practical standpoint the solutions of most interest are those for airfoils having continuously curved surfaces. Curved surfaces require a continuous distribution of line sources or sinks parallel to the generators of the wing surface. As in the analogous development for subsonic speeds, the simplest analytic example is the untapered wing having upper and lower surfaces formed by parabolic arcs. As a starting point, a streamline surface of parabolic shape may be obtained by integrating the solution (15-25) along x. To obtain the integral of Eq. 15-25, we make use of the general relation (Euler's formula)

$$(n + 1)\varphi = xu + yv + zw \qquad (15\text{-}28)$$

valid for any homogeneous potential field of order n in the velocities u, v, and w. For the conical field, $n = 0$ and, since $\varphi = \int u dx$, the right-hand side of Eq. 15-28 yields the integral of the solution (15-25). After introducing the expressions (15-25, 15-26, and 15-24) for u, v, and w there is obtained

$$u_1 = \int u_0 dx = \frac{1}{\pi \sqrt{1 - m^2}} \left[(y - mx) \cosh^{-1} \frac{1 - m\tau}{m - \tau} \right.$$

$$\left. - (y + mx) \cosh^{-1} \frac{1 + m\tau}{m + \tau} - i \sqrt{1 - m^2}\, z \ln \frac{\tau - m}{\tau + m} \right] \qquad (15\text{-}29)$$

Eq. 15-29 yields the horizontal perturbation velocity (and hence the pressure) for a parabolically curved surface whose ordinates are obtained by integrating the ordinates of the wedge along the x direction.

The pressure field for the sweptback wing with biconvex sections may now be obtained by superimposing the solutions (15-25) and (15-29) in the following way: If the root chord of the wing lies between $x = -1$ and $x = +1$ and if we use $u(\pm 1)$ to denote a disturbance field with its origin displaced to the point $x = \pm 1$, then the nose angle of the wing section is obtained by a solution $u_0(-1)$, and the curvature of the surface by $u_1(-1)$. Two additional solutions are required at $x = +1$ to terminate the wing. The resulting disturbance field may be represented by the sum

$$u_0(-1) - u_1(-1) + u_1(+1) + u_0(+1) \qquad (15\text{-}30)$$

This summation yields a relatively simple expression for the pressure coefficient in the plane of the wing:

$$\frac{\Delta p}{\frac{1}{2}\rho_\infty U^2} = \frac{4}{\pi}\frac{t}{c}\frac{m}{\sqrt{1-m^2}}\left[\frac{y'}{m}\left(\cosh^{-1}\frac{x'+1}{|y'-m|} - \cosh^{-1}\frac{x'-1}{|y'+m|}\right.\right.$$

$$\left.\left. + \frac{\bar{y}'}{m}\cosh^{-1}\frac{\bar{x}'-1}{y'-m} - \cosh^{-1}\frac{\bar{x}'+1}{|y'+m|}\right)\right] \quad (15\text{-}31)$$

The pressure distribution over the same wing at subsonic speeds has been given in Art. 9, Eq. 9-19.

Fig. A,15h shows the distribution of pressure as given by Eq. 15-31 at various sections along the span for the case of a wing with 60° sweep at a Mach number $M_\infty = \sqrt{2}$ ($m = 0.577$). These calculations illustrate the nature of the transition from the supersonic, or Ackeret, type of pressure distribution at the root to the subsonic form at sections farther outboard. Comparison with Fig. A,9e shows that the same trends exist in a qualitative sense for the same wing at subsonic speeds. At supersonic speeds, however, the pressure distribution at the root section is identical in form with that given by the Ackeret theory except that the pressures are reduced in the ratio (15-27). This statement holds for the root section of any sweptback wing formed by joining two cylindrical surfaces, provided the root section remains ahead of the zone of influence of the tip sections.

An interesting feature of these calculations is the characteristic sharp compression, or pressure recovery, which occurs along the wave emanating from the trailing edge of the root section. The linearized theory shows an infinite gradient in the pressure behind this wave, but of course in practice the wave will form a shock of finite strength. As in the lifting case (Art. 14), transition to the two-dimensional subsonic type of pressure distribution is essentially complete at the point where this compression wave crosses the leading edge of the wing. At sections beyond that point, the pressure distribution is approximated very closely by the velocity function 20 of Table A,2 after correction for the transverse velocity.

Spanwise distribution of wave drag of sweptback wings. It is apparent from the form of the pressure distributions that the drag of the untapered sweptback wing is concentrated near the root sections, or the foremost sections. The variation of the drag of the individual sections with distance from the root is shown in Fig. A,15i. In the case shown ($m = 0.577$) the drag of the root section is about half the value for a similar section of a straight wing. The section drag falls to zero at the point where the compression wave from the trailing edge crosses the line of maximum thickness. Sections beyond this point have essentially zero pressure drag, so that additions to the span of a wing beyond this point can be made without additional pressure drag.

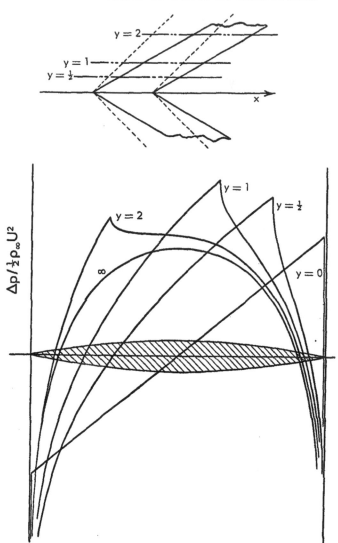

Fig. A,15h. Distribution of pressure at different sections along the span of infinite wing; $M_\infty = 1.4$; $\Lambda = 60°$; biconvex sections.

Although the discussion has been based on the results for an infinite wing, the effect of cutting the wing off at some point to form a tip can be foreseen from the forms of the elements used in the superposition. Thus it can be seen that the pressure distribution at the extreme tip will be composed of two curves (assuming that the wing is long enough to avoid multiple interaction between root and tip), one of which will be the pressure on the infinite wing and the other equal to one half the pressure of the root section but with the sign reversed. The extreme tip section thus has a negative drag, although it may be shown [*190*] that the total drag increment associated with the tip region is zero for an untapered wing.

Fig. A,15j, taken from [*190*], shows the variation of section drag coefficient along the span for three different values of the Mach number

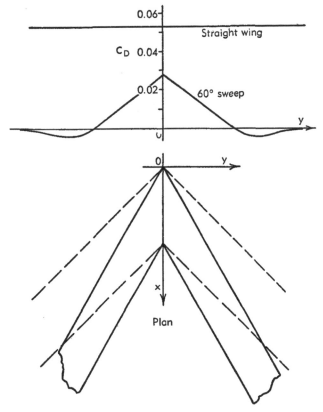

Fig. A,15i. Spanwise distribution of drag of straight and sweptback wings, 10 per cent thick biconvex sections.

in the case of a wing with biconvex sections and 45 degrees of sweep. At the lowest value of the Mach number ($M_\infty = 1.1$) the wing is well inside the Mach cone and the pressure drag falls to zero rapidly on proceeding away from the center section. At $M_\infty = 1.34$ the Mach line lies close to the leading edge and the drag is then nearly constant over the portion of the wing shown. At $M_\infty = 1.41$ the Mach lines coincide with the genera-tors of the wing surface, corresponding to a transverse Mach number of 1. In this case the section drag coefficient approaches infinity as the flow approaches the two-dimensional form at a great distance from the root section. As illustrated by the figure, the actual rate of increase is quite slow, the drag coefficient being ultimately proportional to the square root of the distance from the apex of the wing.

Influence of section shape on the wave drag of swept wings. Since the wave drag of a long narrow wing, swept behind the Mach cone, is concentrated in the vicinity of the root section, von Kármán [43] has proposed that the wave drag of an infinite swept wing be used as an approximation to the drag of a long narrow wing of finite dimensions.

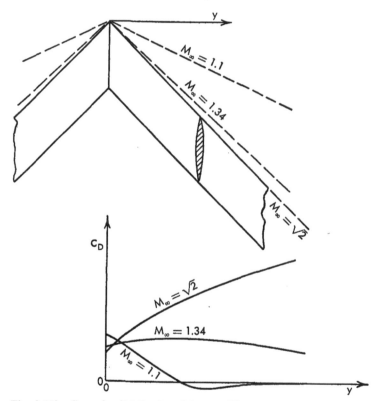

Fig. A,15j. Spanwise distribution of drag at different Mach numbers [190].

Assuming the wing to consist of two cylindrical pieces joined at the root section with the sweepback angle Λ, we find that the equivalent linear source distribution $\sigma'(X, \theta)$ (see Art. 15) has the same form as the distribution of thickness along any streamwise section (see Fig. A,15k). While remaining similar to the section shape, the amplitude of the equivalent source distribution varies with θ in a manner dependent on the sweep angle.

In this case the coefficients A_n in Eq. 15-8 will maintain the same ratio at all values of θ and the drag can be expressed directly in terms of the wing section shape. Supposing a streamwise section of the wing to lie between the points $-c/2$ and $+c/2$ and setting

$$\frac{x}{c/2} = \cos \phi$$

we may write

$$\frac{t(x)}{t(0)} = \sum a_n \sin n\phi \qquad (15\text{-}32)$$

where $t(x)$ is twice the ordinate of the (symmetrical) wing section at x. Eq. 15-10 yields for the drag

$$D = \pi\tfrac{1}{2}\rho_\infty U^2[t(0)]^2 f(M_\infty, \Lambda) \sum_1^\infty na_n^2 \qquad (15\text{-}33)$$

where

$$f(M_\infty, \Lambda) = \frac{\cos^2 \Lambda}{\sin \Lambda} \frac{1 + \sin^2 \Lambda - M_\infty^2 \cos^2 \Lambda}{(1 - M_\infty^2 \cos^2 \Lambda)^{\frac{3}{2}}} \qquad (15\text{-}34)$$

Evaluation of Eq. 15-10 for the root section of the sweptback wing alone yields half the value given above. However, as pointed out by Hayes

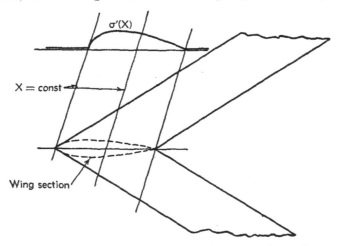

Fig. A,15k.   Equivalent source distribution for sweptback wing.

[44] an allowance must be made for the tips of the wing in obtaining the actual pressure drag. It appears that a suitable approximation to the drag of a wing having a large but finite aspect ratio is obtained when the contribution of the tips to the drag is made equal to the contribution from the root section.

Fig. A,15l shows the drag as calculated by exact formulas from the linearized theory [190] compared to the values given by the approximate equation (15-33). The approximation is good for wings having a sufficiently high aspect ratio and lying well inside the Mach cone from the apex.

Eq. 15-33 is of interest in that it shows the effect of section shape on the drag of wings swept behind the Mach angle. As remarked by Germain

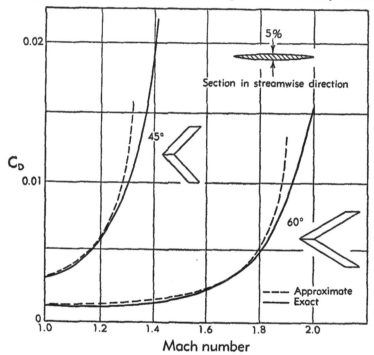

Fig. A,151. Comparison of exact and approximate formulas
for the wave drag of swept wings.

in [*144*], the minimum drag for a given area of the section occurs when the section shape is elliptical. In this case Eq. 15-32 reduces to

$$\frac{t(x)}{t(0)} = \sin \phi \qquad (15\text{-}35)$$

so that the coefficient a_1 is unity and the remaining coefficients are zero.

A,16. Cited References and Bibliography.

Cited References

1. Lamb, H. *Hydrodynamics*, 6th ed. Dover, 1945.
2. Kutta, W. Auftriebskräfte in strömenden Flüssigkeiten. *Illustrierte Aeronaut. Mitt.*, July 1902.
3. Joukowski, N. De la chute dans l'aire de corps légers de forme allongée, animés d'un mouvement rotatoire. *Bull. de l'Inst. Aeronaut. Koutchino, Fasc. I*, St. Petersburg, 1906.

4. Jacobs, E. N. Preliminary report on laminar flow airfoils and new methods adopted for airfoil and boundary layer investigations. *NACA Advance Confid. Rept.*, June 1939.
5. Abbott, I. H., Von Doenhoff, A. E., and Stivers, L. S., Jr. Summary of airfoil data. *NACA Rept. 824*, 1945.
6. Lanchester, F. W. *Aerodynamics.* London, 1907.
7. Munk, M. M. The minimum induced drag of aerofoils. *NACA Rept. 121*, 1921.
8. Prandtl, L., and Betz, A. *Vier Abhandl. zur Hydrodynamik und Aerodynamik,* Göttingen, 1927. Reprinted by Edwards Brothers, 1943.
9. Munk, M. M. *Fluid Dynamics for Aircraft Designers.* Ronald Press, 1929.
10. Ackeret, J. Air forces on airfoils moving faster than sound. *NACA Tech. Mem. 317*, 1925.
11. McCullough, G. B., and Gault, D. E. Boundary-layer and stalling characteristics of the NACA 64A006 airfoil section. *NACA Tech. Note 1923*, 1949.
12. Liepmann, H. W., and Puckett, A. E. *Introduction to Aerodynamics of a Compressible Fluid.* Wiley, 1947.
13. Busemann, A. Drag near sonic velocities. *Third Intern. Congr. of Mechanics,* Stockholm, 1930.
14. Prandtl, L. General considerations on the flow of compressible fluids. *NACA Tech. Mem. 805*, 1936.
15. Glauert, H. The effect of compressibility on the lift of an airfoil. *Aeronaut. Research Council Repts. and Mem. 1135*, 1927.
16. Molenbroeck, P. *Arch. Mat. Phys. 9,* 157–195 (1890).
17. Chaplygin, S. A. Gas jets. *NACA Tech. Mem. 1063*, 1944.
18. von Kármán, Th. *J. Aeronaut. Sci. 8,* 337 (1941).
19. Kaplan, C. Effect of compressibility at high subsonic velocities on the lifting force acting on an elliptic cylinder. *NACA Rept. 834*, 1946.
20. Hantzsche, W., and Wendt, H. *Z. angew. Math. u. Mech. 22,* 72–86 (1942). Transl. as *R.T.P. Transl. 2198, British Ministry of Aircraft Production.*
21. von Kármán, Th. *J. Math. and Phys., 26(3),* 182–190 (1947).
22. Guderley, G. On the transition from a transonic potential flow to a flow with shocks. Transl. as *U.S. Air Force Air Materiel Command Tech. Rept. F-TR-2160-ND*, 1947.
23. Oswatitsch, K., and Weighardt, K. Theoretical analysis of stationary potential flows and boundary layers at high speed. *NACA Tech. Mem. 1189*, 1948.
24. Oswatitsch, K. *Z. angew. Math. u. Mech. 30,* 17–24 (1950).
25. Stack, J., and Lindsey, W. F. Characteristics of low-aspect-ratio wings at supercritical Mach numbers. *NACA Rept. 922*, 1949.
26. Spreiter, J. R. Similarity laws for transonic flow about wings of finite span. *NACA Tech. Note 2273*, 1951.
27. Berndt, S. B. Similarity laws for transonic flow around wings of finite aspect ratio. *Roy. Inst. Technol. Stockholm KTH-Aero. TN 14*, 1950.
28. Kantrowitz, A. R. The formation and stability of normal shock waves in channel flows. *NACA Tech. Note 1225*, 1948.
29. Busemann, A. *J. Aeronaut. Sci. 16,* 337–344 (1949).
30. Kuo, Y. H. *J. Aeronaut. Sci. 18,* 1–6 (1951).
31. Feldman, F. K. Untersuchung von Symmetrischen Tragflugelprofilen bei hohen Unterschallgeschwindigkeiten in einem geschlossenen Windkanal. *Mitt. 14, Inst. für Aerodynamik.* A. G. Gebr. Leeman and Co., Zurich, 1948.
32. Busemann, A. Aerodynamischer Auftrieb bei Überschallgeschwindigkeit. *Volta Congr.,* 328–360 (1935). Also *Luftfahrtforschung 12,* 210–220 (1935).
33. Jones, R. T. Wing plan forms for high-speed flight. *NACA Tech. 863*, 1947.
34. Bateman, H. *Partial Differential Equations.* Dover, 1944.
35. Mathews, C. W., and Thompson, J. R. Comparative drag measurements at transonic speeds of rectangular and sweptback NACA 65-009 airfoils mounted on a freely falling body. *NACA Tech. Note 1969*, 1949.
36. Busemann, A. Pfeilflügel bei Hochgeschwindigkeit. *Lilienthal-Gesellschaft, Bericht 164*, 1943.

37. Struminsky, V. V. *Doklady Akad. Nauk. S.S.S.R. 54*, 765–768 (1946).
38. Jones, R. T. Effects of sweepback on boundary layer and separation. *NACA Rept. 884*, 1947.
39. Sears, W. R. *J. Aeronaut. Sci. 15*, 49 (1948).
40. Relf, E. H., and Powell, C. H. Tests on smooth and stranded wires inclined to the wind direction and a comparison of the results on stranded wires in air and water. *Aeronaut. Research Council Repts. and Mem. 307*, 1917.
41. Kolbe, C. D., and Boltz, F. W. The forces and pressure distribution at subsonic speeds on a plane wing having 45° of sweepback, an aspect ratio of 3, and a taper ratio of 0.5. *NACA Research Mem. A51G31*, 1951.
42. Göthert, B. Ebene und räumliche Strömung bei hohen Unterschallgeschwindigkeiten (Erweiterung der Prandtlschen Regel). *Lilienthal-Gesellschaft, Bericht 127*, 97–101 (1940); *NACA Tech Mem. 1105*, 1946.
43. von Kármán, Th. *J. Aeronaut. Sci. 14*, 373–409 (1947).
44. Hayes, W. D. Linearized supersonic flow. *North Amer. Aviation Co., Rept. AL-222*, Los Angeles, 1947.
45. Munk, M. M. *J. Appl. Phys. 21*, 159–161 (1950).
46. Brown, C. E. The reversibility theorem for thin airfoils in subsonic and supersonic flow. *NACA Tech. Note 1944*, 1949.
47. Ursell, F., and Ward, G. N. *Quart. J. Mech. and Appl. Math. 3*, 326–348 (1950).
48. Heaslet, M. A., and Spreiter, J. R. Reciprocity relations in aerodynamics. *NACA Rept. 1119*, 1953.
49. Flax, A. H. *J. Aeronaut. Sci. 19*, 361–374 (1952).
50. Krienes, K. The elliptic wing based on the potential theory. *NACA Tech. Mem. 971*, 1941.
51. Kinner, W. *Ing.-Arch. 8*, 47 (1937).
52. Whittaker, E. T., and Watson, G. N. *A Course of Modern Analysis*, 4th ed. Cambridge Univ. Press, 1940.
53. Hobson, E. W. *The Theory of Spherical and Ellipsoidal Harmonics*. Cambridge Univ. Press, 1931.
54. Jones, R. T. Correction of the lifting-line theory for the effect of the chord. *NACA Tech. Note 817*, 1941.
55. Lampert, S. *J. Aeronaut. Sci. 18*, 107–114 (1951).
56. Lampert, S. Some applications of conical-flow methods to subsonic lifting-surface problems. *NACA Tech. Note 2262*, 1950.
57. Busemann, A. *Jahrbuch deut. Luftfahrtforschung 7B(3)*, 105–121 (1943); *NACA Tech. Mem. 1100*, 1947.
58. Cohen, D. A method for determining the camber and twist of a surface to support a given distribution of lift, with applications to the load over a sweptback wing. *NACA Rept. 826*, 1945.
59. Falkner, V. M., and Lehrian, D. Low-speed measurements of the pressure distribution at the surface of a swept-back wing. *Aeronaut. Research Council Repts. and Mem. 2741*, 1953.
60. Swanson, R. S., and Crandall, S. M. An electromagnetic analogy method of solving lifting-surface-theory problems. *NACA Advance Restricted Rept. L5D23*, 1945. (Issued as Wartime Rept.)
61. Swanson, R. S., and Priddy, E. L. Lifting-surface-theory values of the damping in roll and of the parameters used in estimating aileron stick forces. *NACA Advance Restricted Rept. L5F23*, 1945. (Issued as Wartime Rept.)
62. Malavard, L., and Duquenne, R. *Recherche Aéronaut. 23*, 3–22 (1951).
63. Allen, D. N. deG., and Dennis, S. C. R. *Quart. J. Mech. and Appl. Math. 4*, 199–208 (1951); *6*, 81–100 (1953).
64. Tyler, C. M., Jr. *Solution of the Lifting-line and Lifting-Surface Integral Equations by the Relaxation Method*. Ph. D. Thesis, Univ. Pittsburgh, 1949.
65. Birnbaum, W. *Z. angew. Math. u. Mech. 3*, 290–297 (1923).
66. Blenk, H. *Z. angew Math. u. Mech. 5*, 36–47 (1925); *NACA Tech. Mem. 1111*, 1947.
67. Jones, W. P. Theoretical determination of the pressure distribution on a finite wing in steady motion. *Aeronaut. Research Council Repts. and Mem. 2145*, 1943.

68. Multhopp, H.   Methods for calculating the lift distribution of wings (subsonic lifting surface theory). *Roy. Aircraft Establishment Rept. Aero. 2353*, 1950.
69. Curtis, A. R.   Tables of Multhopp's influence functions. *Brit. Natl. Phys. Lab. Math. Div. Rept. MA/21/0505*, 1952.
70. Multhopp, H.   *Luftfahrtforschung 15*, 153–169 (1938).
71. Garner, H. C.   Methods of approaching an accurate three-dimensional potential solution for a wing. *Aeronaut. Research Council Rept. 12,222*, 1948.
72. Küchemann, D.   *Aeronaut. Quart. 4*, 261–278 (1953).
73. Garner, H. C.   Swept-wing loading. A critical comparison of four subsonic vortex sheet theories. *Aeronaut. Research Council, C.P. 102*, 1952.
74. Küchemann, D.   A simple method for calculating the span loadings on thin swept wings. *Roy. Aircraft Establishment Rept. Aero. 2392*, 1950.
75. Pistolesi, E.   Betrachtungen über die gegenseitige Beeinflussung von Tragflügelsystemen. *Lilienthal-Gesellschaft*, 214–219, 1937.
76. Mutterperl, W.   The calculation of span load distributions on swept-back wings. *NACA Tech. Note 834*, 1941.
77. Weissinger, J.   Ueber die Auftriebsverteilung von Pfeilflugeln. *FB 1553*, Berlin-Adlershof, 1942; *NACA Tech. Mem. 1120*, 1947.
78. De Young, J., and Harper, C. W.   Theoretical symmetric span loading at subsonic speeds for wings having arbitrary plan form. *NACA Rept. 921*, 1948.
79. Schneider, W. C.   A comparison of the spanwise loading calculated by various methods with experimental loadings obtained on a 45° sweptback wing of aspect ratio 8 at a Reynolds number of 4.0 × 10⁶. *NACA Rept. 1208*, 1954.
80. Weighardt, K.   *Z. angew. Math. u. Mech. 19*, 257 (1939); *NACA Tech. Mem. 963*, 1940.
81. Schlichting, H., and Kahlert, W.   Calculation of lift distribution of swept wings. *Roy. Aircraft Establishment Rept. Aero. 2297*, 1948.
82. Scholz, N.   *Ing.-Arch. 18*, 84–105 (1950).
83. Byrd, P. F.   *Ing.-Arch. 19*, 321–323 (1951).
84. Holme, O.   On the approximate solution of the lifting-surface problem with the aid of discrete vortices. *Roy. Inst. Technol. Stockholm, KTH-Aero TN 6*, 1949.
85. Thwaites, B.   A continuous vortex line method for the calculation of lift on wings of arbitrary plan form. *Aeronaut. Research Council Rept. 12,082*, 1949; Addendum, *Aeronaut. Research Council Rept. 12,389*, 1949.
86. Falkner, V. M.   The solution of lifting plane problems by vortex lattice theory. *Aeronaut. Research Council Repts. and Mem. 2591*, 1947.
87. Falkner, V. M.   Calculated loadings due to incidence of a number of straight and swept-back wings. *Aeronaut. Research Council Repts. and Mem. 2596*, 1948.
88. Falkner, V. M.   *Aircraft Eng. 22*, 296 (1950).
89. Staff of the Mathematics Division, N. P. L.   Tables of complete downwash due to a rectangular vortex. *Aeronaut. Research Council Repts. and Mem. 2461*, 1953. Lehrian, D. E.   A calculation of the complete downwash in three dimensions due to a rectangular vortex. *Aeronaut. Research Council Repts. and Mem. 2771*, 1953.
90. Holme, O. A. M.   Measurements of the pressure distribution on rectangular wings of different aspect ratios. *FFA Meddelande 37*, Stockholm-Elvsunda, 1950.
91. Jacobs, W.   *Ing.-Arch. 18*, 344 (1950).
92. Letko, W., and Goodman, A.   Preliminary wind-tunnel investigation at low speed of stability and control characteristics of swept-back wings. *NACA Tech. Note 1046*, 1946.
93. Winter, H.   Flow phenomena on plates and airfoils of short span. *NACA Tech. Mem. 798*, 1936.
94. Jacobs, E. N., and Clay, W. C.   Characteristics of the NACA 23012 airfoil from tests in the full-scale and variable-density tunnels. *NACA Rept. 530*, 1935.
95. Goett, H. J., and Bullivant, W. K.   Tests of NACA 0009, 0012, and 0018 airfoils in the full-scale tunnel. *NACA Rept. 647*, 1939.
96. Zimmerman, C. H.   Characteristics of Clark Y airfoils of small aspect ratios. *NACA Rept. 431*, 1932.
97. Goodman, A., and Brewer, J. D.   Investigation at low speeds of the effect of

aspect ratio and sweep on static and yawing stability derivatives of untapered wings. *NACA Tech. Note 1669*, 1948.

98. Jacobs, E. N., Ward, K. E., and Pinkerton, R. M. The characteristics of 78 related airfoil sections from tests in the variable-density wind tunnel. *NACA Rept. 460*, 1933.

99. Lange, and Wacke. Pruefbericht über 3- und 6- Komponenten-messungen an der Zuspitzungsreihe von Fluegeln kleiner Streckung. *UM 1023, Berlin-Adlershof*, 1943. (Partial report on triangular wings translated as *NACA Tech. Mem. 1176*, 1948.)

100. Wick, B. H. Chordwise and spanwise loadings measured at low speed on a triangular wing having an aspect ratio of two and an NACA 0012 airfoil section. *NACA Tech. Note 1650*, 1948.

101. Staff of Navy Aerodynamical Lab. Air force and moment for gliding wing. *Department of Aeronautics Rept. 677*, 1943.

102. Berndt, S. B. Three component measurement and flow investigation of plane delta wings at low speeds and zero yaw. *Roy. Inst. Technol. Stockholm KTH-Aero TN 4*, 1948.

103. Graham, D. Chordwise and spanwise loadings measured at low speeds on a large triangular wing having an aspect ratio of 2 and a thin, subsonic-type airfoil section. *NACA Research Mem. A50A04a*, 1950.

104. Anderson, A. Chordwise and spanwise loadings measured at low speed on large triangular wings. *NACA Research Mem. A9B17*, 1949.

105. Jones, R. T. Properties of low-aspect-ratio pointed wings at speeds below and above the speed of sound. *NACA Rept. 835*, 1946.

106. Munk, M. M. The aerodynamic forces on airship hulls. *NACA Rept. 184*, 1924.

107. Tsien, H. S. *J. Aeronaut. Sci. 5*, 480 (1938).

108. Lawrence, H. R. The lift distribution on low-aspect-ratio wings at subsonic speeds. *Inst. Aeronaut. Sci. Preprint 313*, 1951.

109. Adams, M. C., and Sears, W. R. *J. Aeronaut. Sci. 20*, 85–98 (1953).

110. Lomax, H., and Heaslet, M. A. Linearized lifting-surface theory for swept-back wings with slender plan forms. *NACA Tech. Note 1992*, 1949.

111. Robinson, A. *Aeronaut. Quart. 4*, 69–82 (1952).

112. Mangler, K. W. Calculation of the pressure distribution over a wing at sonic speeds. *Roy. Aircraft Establishment Rept. Aero. 2439*, 1951.

113. Legendre, R., Eichelbrenner, E. A., and von Baranoff, A. Écoulement transsonique autour d'ailes a forte flèche. *Office natl. Études et Recherches aéronaut 53*, 1952.

114. Mirels, H. Aerodynamics of slender wings and wing-body combinations having swept trailing edges. *NACA Tech. Note 3105*, 1954.

115. Eichelbrenner, E. A. Théorie des corps élancés. *Office natl. Études et Recherches aéronaut. 68*, 1954.

116. Heaslet, M. A., Lomax, H., and Spreiter, J. R. Linearized compressible-flow theory for sonic flight speeds. *NACA Rept. 956*, 1950.

117. Robinson, A., and Young, A. D. Note on the application of the linearized theory for compressible flow to transonic speeds. *College of Aeronautics, Cranfield, Rept. 2*, 1947.

118. Bollay, W. *Z. angew. Math. u. Mech. 19*, 21-35 (1939).

119. Kriesis, P. *Z. angew. Math. u. Mech. 24*, 1-5 (1944).

120. Flax, A. H., and Lawrence, H. R. The aerodynamics of low-aspect-ratio wings and wing-body combinations. *Third Anglo-American Aeronaut. Conf.*, 1951.

121. Durand, W. F., ed. *Aerodynamic Theory*. Springer, Berlin, 1934-1936.

122. Zahm, A. F. Flow and force equations for a body revolving in a fluid. *NACA Rept. 323*, 1929.

123. Hess, R. V., and Gardner, C. S. Study by the Prandtl-Glauert method of compressibility effects and critical Mach number for ellipsoids of various aspect ratio and thickness ratios. *NACA Tech. Note 1792*, 1949.

124. Jones, R. T. Subsonic flow over thin oblique airfoils at zero lift. *NACA Tech. Note 1340*, 1947. Republished as *NACA Rept. 902*, 1948.

125. Neumark, S. Velocity distribution on straight and swept-back wings of small thickness and infinite aspect ratio at zero incidence. *Roy. Aircraft Establishment Rept. Aero. 2200. ARC 10907*, 1947.

126. Küchemann, D. Wing junction, fuselage and nacelles for swept-back wings. *Roy Aircraft Establishment Rept. Aero. 2219. ARC 11035*, 1947.

127. Jones, R. T. Thin oblique airfoils at supersonic speed. *NACA Tech. Rept. 851*, 1946.

128. Busemann, A. *Schriften. deut. Akad. für Luftfahrtforschung 30*, 17–36 (1940). Transl. as *R.T.P. Trans. 1830, British Ministry of Aircraft Production.*

129. Heaviside, O. *Electromagnetic Theory.* Dover, 1950.

130. Epstein, P. S. *Proc. Natl. Acad. Sci. 17*, 532–547 (1931).

131. Hugoniot, H. *J. École Polytech. 57*, 3–97 (1887).

132. Meyer, T. The two-dimensional phenomena of motion in a gas flowing at supersonic velocity. *Dissertation*, Göttingen, 1908. (Reprinted in Carrier, G. F. *Foundations of High Speed Aerodynamics.* Dover, 1951.)

133. Ferri, A. *Elements of Aerodynamics of Supersonic Flows.* Macmillan, 1949.

134. Ferri, A. Investigations and experiments in the Guidonia supersonic wind tunnel. *NACA Tech. Mem. 901*, 1939.

135. Busemann, A., and Walchner, O. *Forsch. Gebiete Ingenieurw. 4A*, 87–92 (1933).

136. Sanger, E. *Raketen-Flugtechnik.* Oldenbourg, Munich and Berlin, 1933.

137. Chapman, D. R. Airfoil profiles for minimum pressure drag at supersonic velocities—General analysis with application to linearized supersonic flow. *NACA Tech. Rept. 1063*, 1952.

138. Chapman, D. R. Airfoil profiles for minimum pressure drag at supersonic velocities—Application of shock-expansion theory including consideration of the hypersonic range. *NACA Tech. Note 2787*, 1952.

139. Bergman, S. On solutions with algebraic character of linear partial differential equations. *Trans. Am. Math. Soc. 68*, 461–507 (1950).

140. Glauert, H. *Aerofoil and Airscrew Theory.* Macmillan, 1943.

141. Mirels, H. Lift-cancellation technique in linearized supersonic wing theory. *NACA Rept. 1004*, 1950.

142. Lagerstrom, P. A. Linearized supersonic theory of conical wings. *NACA Tech. Note 1685*, 1948.

143. Hayes, W. D., Browne, S. H., and Lew, R. J. Linearized theory of conical supersonic flow with application to triangular wings. *North Amer. Aviation Rept. NA-46-818*, 1946.

144. Germain, P. La théorie générale des mouvements coniques et ses applications à l'aérodynamique supersoniques. *Office natl. Études et Recherches aéronaut. 34*, 1949. Transl. as *NACA Tech. Mem. 1354*, 1955.

145. Robinson, A. Aerofoil theory of a flat delta wing at supersonic speeds. *Brit. Aeronaut. Research Council Repts. and Mem. 2548*, 1946.

146. Squire, H. B. An example in wing theory at supersonic speed. *Aeronaut. Research Council Repts. and Mem. 2549*, 1951.

147. Multhopp, H. A unified theory of supersonic wing flow, employing conical fields. *Roy. Aircraft Establishment Rept. Aero. 2415*, 1951.

148. Ribner, H. S. Some conical and quasi-conical flows in linearized supersonic wing theory. *NACA Tech. Note 2147*, 1950.

149. Walker, H. J., and Ballantyne, M. B. Pressure distribution and damping in steady roll at supersonic Mach numbers of flat sweptback wings with subsonic edges. *NACA Tech. Note 2047*, 1950.

150. Walker, H. J., and Ballantyne, M. B. Pressure distribution and damping in steady pitch at supersonic Mach numbers of flat sweptback wings having all edges subsonic. *NACA Tech. Note 2197*, 1950.

151. Byrd, P. F., and Friedman, M. D. *Handbook of Elliptic Integrals for Engineers and Physicists.* Grundlehren der Mathematischen Wissenschaften, Band LXVIII. Springer-Verlag, Heidelberg, 1953.

152. Baldwin, B. S., Jr. Triangular wings cambered and twisted to support specified distributions of lift at supersonic speeds. *NACA Tech. Note 1816*, 1949.

153. Bolton-Shaw, B. W. Nose controls on delta wings, at supersonic speeds. *College of Aeronautics, Cranfield, Rept. 36*, 1950.
154. Lagerstrom, P. A., and Graham, M. E. Downwash and sidewash induced by three-dimensional lifting wings in supersonic flow. *Douglas Aircraft Co. Rept. SM13007*, Apr. 1947.
155. Cohen, D. Formulas for the supersonic loading, lift, and drag of flat sweptback wings with leading edges behind the Mach lines. *NACA Rept. 1050*, 1951.
156. Lagerstrom, P. A., and Graham, M. E. Linearized theory of supersonic control surfaces. *Douglas Aircraft Co. Rept. SM13060*, 1947.
157. Frick, C. W., Jr. Application of the linearized theory of supersonic flow to the estimation of control-surface characteristics. *NACA Tech. Note 1554*, 1948.
158. Cohen, D., and Friedman, M. D. Theoretical investigation of the supersonic lift and drag of thin, sweptback wings with increased sweep near the root. *NACA Tech. Note 2959*, 1953.
159. Puckett, A. E. *J. Aeronaut. Sci. 13*, 475–484 (1946).
160. Evvard, J. C. Use of source distributions for evaluating theoretical aerodynamics of thin finite wings at supersonic speeds. *NACA Rept. 951*, 1950.
161. Krasilshchikova, E. A. *Uchenye Zapiski 154, Mekhanika 4*, 181–239 (1951). Transl. *NACA Tech. Mem. 1383*, 1956.
162. Goodman, T. R. The lift distribution on non-conical flow regions of thin finite wings in a supersonic stream. *Cornell Aeronaut. Lab. Rept.*, Oct. 1948.
The problem of edges in the doublet distribution method of obtaining supersonic lift. *Cornell Aeronaut. Lab. Rept.*, Feb. 1950.
163. Heumann, C. *J. Math. and Phys. 20*, 127–206 (1941).
164. Behrbohm, H., and Oswatitsch, K. *Ing.-Arch. 18*, 370–377 (1950).
165. Cramer, R. H. Interference between wing and body at supersonic speeds, Part V, Phase 1. *Cornell Aeronaut. Lab. Rept. CAL/CM 597*, 1950.
166. Coleman, T. F. Supersonic lift solutions obtained by extending the simple linearized conical flow theory. *North Amer. Aviation Co. Rept. CM440*, 1948.
167. Behrbohm, H. The lifting trapezoidal wing with small aspect ratio at supersonic speed. *SAAB Aircraft Co. Tech. Note 10*, Linköping, Sweden, 1952.
168. Lagerstrom, P. A., and Graham, M. E. Low aspect ratio rectangular wings in supersonic flow. *Douglas Aircraft Co. Rept. SM15110*, 1947.
169. Picard, C., and Chevalier, J. P. *Recherche aéronaut. 26*, 3–12 (1952).
170. Nielsen, J. N., Matteson, F. H., and Vincenti, W. G. Investigation of wing characteristics at a Mach number of 1.53. III: Unswept wings of differing aspect ratio and taper ratio. *NACA Research Mem. A8E06*, 1948.
171. Dye, F. E., Jr. Interference between wing and body at supersonic speeds. Part IX: Pressure distribution tests of wing-body interference models at Mach No. of 2.0. Phase V tests of Aug. 1951. *Cornell Aeronaut. Lab. Rept. CAL/CF 1684*, Dec. 1951.
172. Puckett, A. E., and Stewart, H. J. *J. Aeronaut. Sci. 14*, 567–578 (1947).
173. Drougge, G., and Larsson, P. O. *Report of the Aeronautical Research Inst. of Sweden (FFA)*. To be published.
174. Ellis, M. C., Jr., and Hasel, L. E. Preliminary tests at supersonic speeds of triangular and swept-back wings. *NACA Research Mem. 26L17*, 1947.
175. Drougge, G. Some measurements at low supersonic speeds by a method for continuous variation of the Mach number. *FFA Rept. 42*, Stockholm-Elvsunda, 1952.
176. Vincenti, W. G., Nielson, J. N., and Matteson, F. H. Investigation of wing characteristics at a Mach number of 1.53. I: Triangular wings of aspect ratio 2. *NACA Research Mem. A7I10*, 1947.
177. Chang, C. C., and Houser, J. E. Correlation of test data and comparison with theoretical results for thin wings in supersonic flow. *Glenn L. Martin Co. Eng. Rept. 2729*, 1947.
178. Harmon, S. M., and Jeffreys, I. Theoretical lift and damping in roll of thin wings with arbitrary sweep and taper at supersonic speeds. Supersonic leading and trailing edges. *NACA Tech. Note 2114*, 1950.

179. Gilles, A. *Recherche aéronaut. 19*, 3-10 (1951).
180. Lapin, E. Charts for the computation of lift and drag of finite wings at supersonic speeds. *Douglas Aircraft Co. Rept. SM-13480*, 1949.
181. Martin, J. C., and Jeffreys, I. Span load distributions resulting from angle of attack, rolling and pitching for tapered swept-back wings with streamwise tips. Supersonic leading and trailing edges. *NACA Tech. Note 2643*, 1952.
182. Hannah, M. E., and Margolis, K. Span load distributions resulting from constant angle of attack, steady rolling velocity, steady pitching velocity, and constant vertical acceleration for tapered sweptback wings with streamwise tips. Subsonic leading edges and supersonic trailing edges. *NACA Tech. Note 2831*, 1952.
183. Eichelbrenner, E. A. *Recherche aéronaut. 25*, 19-20 (1952).
184. Oswatitsch, K. Die theoretischen Arbeiten über schallnahe Strömung am Flugtechnischen Inst. der Kungl. Tekniska Högskolan, Stockholm. *Proc. Eighth Intern. Congress on Theoretical and Appl. Mech.*, 1953.
185. Ward, G. N. *Quart. J. Mech. and Appl. Math. 2*, 75-97 (1949).
186. Whitcomb, R. T. A study of the zero-lift drag-rise characteristics of wing-body combinations near the speed of sound. *NACA Research Mem. L52H08*, 1952.
187. Jones, R. T. *J. Aeronaut. Sci. 19*, 813-822 (1952).
188. Sears, W. R. *Quart. Appl. Math. 5*, 361-366 (1947).
189. Haack, W. Geschossformen kleinsten Wellenwiderstandes. Lilienthal-Gesellschaft, *Bericht 139*, 1941.
190. Harmon, S. M., and Swanson, M. D. Calculation of the supersonic wave drag of non-lifting wings with arbitrary sweepback and aspect ratio. *NACA Tech. Note 1319*, 1947.
191. Harmon, S. M. Theoretical supersonic wave drag of untapered sweptback and rectangular wings at zero lift. *NACA Tech. Note 1449*, 1947.
192. Bismut, M. *Recherche aéronaut. 21*, 9-18 (1951).
193. Chang, C. C. Applications of von Kármán's integral method in supersonic wing theory. *NACA Tech. Note 2317*, 1951.
194. Nielsen, J. N. The effect of aspect ratio and taper on the pressure drag at supersonic speeds of unswept wings at zero lift. *NACA Tech. Note 1487*, 1947.
195. Margolis, K. Supersonic wave drag of sweptback tapered wings at zero lift. *NACA Tech. Note 1448*, 1947.
196. Margolis, K. Effect of chordwise location of maximum thickness on the supersonic wave drag of sweptback wings. *NACA Tech. Note 1543*, 1948.
197. Margolis, K. Supersonic wave drag of nonlifting sweptback tapered wing with Mach lines behind the line of maximum thickness. *NACA Tech. Note 1672*, 1948.
198. Lawrence, T. Charts of the wave drag of wings at zero lift. *Roy. Aircraft Establishment Tech. Note Aero. 2139*, 1952.
199. Havelock, T. H. Calculations illustrating the effect of boundary layer on wave resistance. *Trans. Inst. Nav. Arch. 90*, 259-271 (1948).
200. von Kármán, Th. The problems of resistance in compressible fluids. *Proc. 5th Volta Congr.* 255-64 (1935).
201. von Kármán, Th., and Moore, N. B. *Trans. Am. Soc. Mech. Engrs. 54*, 303-310 (1932).

Bibliography for Table A,2

Allen, H. J. General theory of airfoil sections having arbitrary shape or pressure distribution. *NACA Rept. 833*, 1945.
Glauert, H. *The Elements of Aerofoil and Airscrew Theory.* Cambridge Univ. Press, 1943.
Munk, M. M. *Fundamentals of Fluid Dynamics for Aircraft Designers.* Ronald Press, 1929.